Big Data Analytics for Cyber-Physical Systems

Shiyan Hu • Bei Yu

Editors

Big Data Analytics for Cyber-Physical Systems

 Springer

Editors
Shiyan Hu
School of Electronics
and Computer Science
University of Southampton
Southampton, UK

Bei Yu
Department of Computer Science
and Engineering
The Chinese University of Hong Kong
Shatin, Hong Kong

ISBN 978-3-030-43496-0   ISBN 978-3-030-43494-6   (eBook)
https://doi.org/10.1007/978-3-030-43494-6

This Springer imprint is published by the registered company Springer Nature Switzerland AG
The registered company address is: Gewerbestrasse 11, 6330 Cham, Switzerland

Preface

This book highlights research and survey articles dedicated to big data techniques for cyber-physical systems, which address the close interactions and feedback controls between cyber components and physical components. This book discusses fundamental big data problems and solutions in large-scale distributed cyber-physical systems. It then addresses the design and control challenges in multiple application domains such as vehicular system, smart city, smart building, and digital microfluidic biochips. It also discusses the recent advances and trends in the maritime simulation system and the flood defense system.

Some of the unique features of this book are listed as follows:

- Helps readers understand the fundamentals of how big data analytics and optimization are involved in developing the cyber-physical systems.
- Presents readers with practical tools and design methodologies for implementing highly efficient big data based cyber-physical systems.
- Introduces recent advances and trends in leveraging big data techniques in a wide spectrum of domains of cyber-physical systems.

With these features covered, this book can serve as a useful reference for students and researchers in the field of computer science, computer engineering, and electrical engineering.

We would like to thank our colleagues and all the authors for their insight and contributions. We also thank all the referees for carefully reviewing the chapters and making many suggestions for improvements. Last but not the least, we would like to thank Springer editors Murugesan Tamilselvan and Mary James for their continuous and efficient support.

Southampton, UK   Shiyan Hu
Hong Kong, China   Bei Yu

Contents

Contributors

Jahan Ali Faculty of Computer Science and Software Engineering, University Malaysia Pahang, Pahang, Malaysia

A. B. M. Alim Al Islam Department of CSE, Bangladesh University of Engineering and Technology, Dhaka, Bangladesh

A. Taufiq Asyhari Centre for Electronic Warfare, Information and Cyber, Cranfield University, Defence Academy of the UK, Shrivenham, UK

Ansuman Banerjee ACMU, Indian Statistical Institute, Kolkata, India

Nilanjan Banerjee Onometra Technologies Pvt. Ltd., India

Md Zakirul Alam Bhuiyan Department of Computer and Information Sciences, Fordham University, Bronx, NY, USA

Lin Cao College of Petroleum Engineering, Yangtze University, Wuhan, China

Soumi Chattopadhyay Indian Institute of Information Technology, Guwahati, India

Tinghuan Chen CSE Department, Chinese University of Hong Kong, Hong Kong, China

Xiaodao Chen School of Computer Science, China University of Geosciences, Wuhan, China

Pubali Datta Department of Computer Science, University of Illinois Urbana Champaign, Champaign, IL, USA

Ze Deng School of Computer Science, China University of Geosciences, Wuhan, Hubei, P.R. China
Hubei Key Laboratory of Intelligent Geo-Information Processing, China University of Geosciences (Wuhan), Wuhan, P.R. China

Swarnava Dey TCS Research and Innovation, Kolkata, India

Hao Geng CSE Department, Chinese University of Hong Kong, Hong Kong, China

Qi Guo Wuhan Tianhong Lightning Protection Testing Center Development Co., Ltd., Wuhan, China

Shiyan Hu School of Electronics and Computer Science, University of Southampton, Southampton, UK

Xiaohui Huang School of Computer Science, China University of Geosciences, Wuhan, China

Mohammad Nomani Kabir Faculty of Computer Science and Software Engineering, University Malaysia Pahang, Pahang, Malaysia

Taslim Arefin Khan Department of CSE, Bangladesh University of Engineering and Technology, Dhaka, Bangladesh

Bingqing Lin Shenzhen University, Shenzhen, China

Xiangang Luo School of Geography and Information Engineering, China University of Geosciences, Wuhan, People's Republic of China
National Engineering Research Center for Geographic Information System, Wuhan, China

Lin Mu College of Marine Science and Technology, China University of Geosciences, Wuhan, People's Republic of China

Arijit Mukherjee TCS Research and Innovation, Kolkata, India

Mahmuda Naznin Department of CSE, Bangladesh University of Engineering and Technology, Dhaka, Bangladesh

Xuewei Ning College of Petroleum Engineering, Yangtze University, Wuhan, China

Himadri Sekhar Paul TCS Research and Innovation, Kolkata, India

Md Arafatur Rahman Faculty of Computer Science and Software Engineering, University Malaysia Pahang, Pahang, Malaysia
IBM Center of Excellence, University Malaysia Pahang, Pahang, Malaysia

Suraiya Tairin Department of CSE, Bangladesh University of Engineering and Technology, Dhaka, Bangladesh

Chaowei Wan School of Computer Science, China University of Geosciences, Wuhan, China

Yuewei Wang School of Computer Science, China University of Geosciences, Wuhan, China

Di Wu School of Geography and Information Engineering, China University of Geosciences, Wuhan, People's Republic of China
National Engineering Research Center for Geographic Information System, Wuhan, China

Zhi Xiong Department of Engineering Physics, Tsinghua University, Beijing, P.R. China

Zhanya Xu School of Geography and Information Engineering, China University of Geosciences, Wuhan, People's Republic of China
National Engineering Research Center for Geographic Information System, Wuhan, China

Guannan Yao Department of Engineering Physics, Tsinghua University, Beijing, P.R. China

Bei Yu Department of Computer Science and Engineering, The Chinese University of Hong Kong, Shatin, Hong Kong

Xingkai Zhang College of Petroleum Engineering, Yangtze University, Wuhan, China

Enjin Zhao Shenzhen Research Institute, China University of Geosciences, Shenzhen, People's Republic of China

Hui Zhao College of Petroleum Engineering, Yangtze University, Wuhan, China

Shaobo Zhong Beijing Research Center of Urban Systems Engineering, Beijing, P.R. China

Junlong Zhou School of Computer Science and Engineering, Nanjing University of Science and Technology, Nanjing, China

Shuang Zhu School of Geography and Information Engineering, China University of Geosciences, Wuhan, People's Republic of China
National Engineering Research Center for Geographic Information System, Wuhan, China

Wei Zhu Beijing Research Center of Urban Systems Engineering, Beijing, P.R. China

Chapter 1
Cyber-Physical Systems: An Overview

Bei Yu, Junlong Zhou, and Shiyan Hu

1.1 Introduction

Cyber-physical system (CPS) is a complex and heterogeneous system with seamlessly integrated cyber components (e.g., sensors, computers, control centers, and actuators) and physical processes involving mechanical components, human activities, and surrounding environment [1]. CPS is capable of closely interacting with the surrounding physical environment through perception, communication, computation, and control. There are various CPS application domains including electric vehicle (EV), smart grid, health and medicine, smart home, and advanced industries, which promise substantial economic and social benefits.

The terminology of CPS was coined in 2006, when scientists and engineers from different discipline backgrounds started to work on the different aspects of CPS. System designers leverage scientific methodologies as well as powerful engineering techniques and tools for time or frequency domain analysis, pattern recognition, modern filtering, prediction, estimation, optimization, and machine learning to optimize CPS for high performance. Meanwhile, computer scientists and engineers have made breakthroughs in novel programming languages, powerful

B. Yu
Department of Computer Science and Engineering, The Chinese University of Hong Kong,
Shatin, Hong Kong
e-mail: byu@cse.cuhk.edu.hk

J. Zhou (✉)
School of Computer Science and Engineering, Nanjing University of Science and Technology,
Nanjing, China
e-mail: jlzhou@njust.edu.cn

S. Hu
School of Electronics and Computer Science, University of Southampton, Southampton, UK
e-mail: S.Hu@soton.ac.uk

© Springer Nature Switzerland AG 2020
S. Hu, B. Yu (eds.), *Big Data Analytics for Cyber-Physical Systems*,
https://doi.org/10.1007/978-3-030-43494-6_1

modeling formulations and advanced verification tools, energy-efficient computing techniques, compiler tools, embedded systems and software, and innovative methodologies to improve CPS reliability, efficiency, and security. However, the design and development of CPS face new challenges incurred by the ever increasing system scale and complexity, the interaction with the complex physical world, the adoption of distributed embedded systems, and the stringent requirements on reliability, efficiency, and security. A key method addressing these challenges is to co-design physical and cyber components. For example, CPS design needs to model, simulate, and verify the sensing, calibration, computation, control and communication infrastructures, the software and hardware platform, and the surrounding physical environment, simultaneously.

Although CPS has been studied for more than 10 years, its development is still in an early stage. There are plenty of opportunities and technical challenges in ensuring system reliability, efficiency, and security of multiple domains such as smart grid, smart building, electric vehicle, and healthcare. Below, we first elaborate the opportunities, challenges, and recent work in these CPS application domains, then introduce the solutions to improving system QoS from reliability and security perspectives, and finally present the design methodologies from design automation, data analysis to hardware support.

1.2 Applications

In this section, we give a systematic review of typical CPS applications including smart grid, smart building, electric vehicle, and healthcare.

Smart Grid As shown in Fig. 1.1, smart grid is a system that integrates modern communication technologies, emerging energy storage devices, and renewable energy sources with the traditional electricity grid to facilitate electric energy generation, distribution, and transmission with high efficiency, reliability, and security [2]. In a smart grid, cyber systems and physical systems are closely coupled to address various challenges such as the intermittency of regeneration and the uncertainties in the energy market. A particularly interesting aspect in smart grid CPS is the dynamic electricity pricing mechanism. In reality, utility companies always adopt dynamic pricing mechanisms to balance the electricity load. For example, a common pricing strategy is that the electricity price of peak hours is set to higher than that of valley hours. This strategy would encourage customers to use household appliances during valley hours and hence mitigate the electrical overload at peak hours. On the other hand, with the help of the massively deployed smart metering infrastructures, smart homes can use the dynamic pricing mechanisms to control household appliances such that electricity energy is saved and electricity bill is reduced [3]. Consequently, there will be a large amount of reduction of energy consumption in the entire electricity grid. According to statistics [4], if each resident house saves 5% energy, the reduction in energy consumption across the USA will

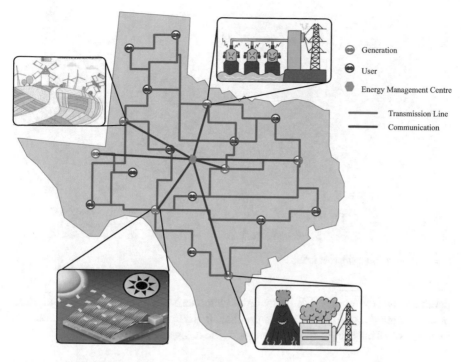

Fig. 1.1 Illustration of smart grid

be approximate to the amount of energy savings achieved by removing 52 million cars. As it is effective in decreasing electricity bill, dynamic pricing mechanism is also widely adopted by cloud service providers to minimize the cost [5, 6].

Smart Building Smart buildings (as shown in Fig. 1.2) play an important role in human daily life since they provide a more comfortable life environment with reduced energy consumption (and hence electricity bill) as compared to traditional buildings [7]. To meet the increasing demand for high comfort level and intelligent building services, heating, ventilation, and air conditioning (HVAC) systems need to be controlled efficiently in smart buildings to reduce energy consumption and provide a good thermal performance [8]. On one hand, simulation is one of the most important steps at the design stage of smart buildings. In this step, energy consumption and thermal performance are predicted to assist energy and thermal-aware optimization. In the simulation, heating and cooling systems, equipment, and building components are fed into the building simulation platform to reduce energy consumption and mitigate temperature using heat balance algorithms [9–11]. On the other hand, implementation of advanced sensor calibration for extending the lifetime of smart buildings is an indispensable step in the design of smart buildings. Different from traditional sensor calibration, the sensing matrix of advanced sensors deployed in smart buildings depends on weather as well as the position of sensors and

Fig. 1.2 Features of smart buildings

parameters of the building (e.g., material characteristics, geometry, and equipment power per area) [9, 10]. However, in practice, it is hard to obtain all of these complex information. This brings new challenges in the design of smart buildings.

Electric Vehicle Electric vehicle (EV), as a green and clean transportation tool, has many advantages in terms of energy conversion efficiency and environmental protection over traditional vehicles. As a CPS, EV also consists of actual physical systems and cyber components for computation, control, and sensing. The two parts work together to make EV energy-efficient and environmental friendly. Due to these advantages, EVs have been widely studied in the literature and deployed in the real-world. However, the increasing penetration level of EV brings new challenges and problems. These potential issues include voltage deviation as well as power quality and overload problems caused by uncoordinated charging of EVs. In particular, massive amounts of stochastic charging to EV have negative effects on the capacity and stability of EV systems and power systems. To overcome these challenges, numerous coordinated charging strategies are proposed in the recent years.

Health and Medicine CPS is widely adopted in medical and healthcare applications to provide health related services. Health and medicine are always considered as one of the most important issues in the world. Many countries are suffering severe shortage of healthcare, which causes an increasing health cost. Currently, a major concern is the ineffectiveness to process the massive data in CPS health and medicine applications. As a result, advanced computing frameworks and infrastructures are necessary.

An innovative revolution in computing and communication networks has been achieved in recent years. As the feature size of semiconductors enters nanometer era, biochips become smaller and smaller [12]. In addition, with the development of near-threshold and low-power integrated circuit design, biochips have higher performance and lower energy consumption [13]. These advanced integrated circuits and semiconductor technologies integrated with CPS bring many benefits to health related applications. For example, advanced biochips facilitate in vivo implantation of human to monitor health condition by interacting with the monitoring center. However, medical CPS faces extensive challenges, such as security and reliability [14]. Different from common computing applications, in order to obtain multi-model information from biological, cognitive, and social networks, various resources with interior sensing abilities are adopted to extend beyond computer-connected physical process [15]. Therefore, some standards are modified to comprise interdependent information networks, smart devices, and mobile devices.

Another typical application of medical and healthcare CPS is biochip manufacturing. Biochips can control the movement of droplets of picoliter volumes. Digital microfluidic biochips are used in biomolecular recognition to perform point-of-care clinical diagnostics. These point-of-care tests can be used in resource-limited areas [16]. With biochips, the reactions of flash chemistry can yield desired substances with high selectivity [17]. The biochips can promote the miniaturization and the automation for the bioassays of samples [18].

1.3 Quality of Service

The purpose of designing and deploying CPS is to provide a high quality of service (QoS) for specific users. In this section, we focus on measuring the QoS of CPS from two representative aspects of reliability and security.

Reliability Typically, the CPS reliability refers to the ability of correctly performing system intended function under a given set of environmental and operational conditions during a given time interval [19]. Recently, Cao et al. [20–22] investigate reliability enhancement algorithms for both mixed-criticality uniprocessor embedded systems and heterogeneous multiprocessor systems. Zhou et al. [23] explore reliability and temperature constrained task scheduling for minimizing system makespan on heterogeneous multiprocessor platforms. Meanwhile, the authors further design a framework to deal with the energy, makespan, and lifetime issues in reliability-driven task scheduling [24]. Haque et al. [25] address the problem of satisfying a specific reliability goal for periodic real time tasks executing on a multiprocessor system with minimum energy consumption. Li et al. [26, 27] propose a set of feedback control-based algorithms for reliability improvement of EtherCAT-based networked systems.

Security The interdependence and interaction between cyber components and the physical world inevitably bring new security risks to CPS. A cyber disturbance, either an occasional component failure or a malicious attack, may cause catastrophic consequences in CPS. For example, the attacks of destroying transmission lines by increasing energy load during peak hours or undermining generators by increasing frequency disturbance are very likely to result in widespread electricity blackouts [28]. Therefore, it is necessary to investigate useful methodologies to guarantee security for CPS. In the area of CPS security, abnormal detection is an important step to identify either short-term or long-term abnormal system behaviors. For short-term abnormal detection, the popular techniques of support vector machine and Gaussian mixed model are shown effective to detect physical attacks, cyberattacks, and data attacks for energy theft in the smart grid [29]. In order to enable long-term abnormal detection, a partially observable Markov decision process is proposed to find the optimal action that maximizes the expected reward or, equivalently, minimizes the long-term impact of cyberattacks in CPS [30].

1.4 Methodologies

In this section, we give a brief introduction to three representative design methodologies in CPS, including design automation, data analytics, and hardware support.

1.4.1 Design Automation

In the design of CPS, multiple factors need to be considered, such as energy consumption, reliability, timing, fault tolerance, and security. Design automation tools can be used to develop high performance CPS through complex analytical modeling, efficient simulation, synthesis, and verification. There have been several instructive attempts (e.g., [9]) to bring CAD methodologies developed for traditional integrated circuit design to deal with the new CPS research challenges, such as how to improve CPS performance subject to performance constraints and how to reduce power consumption in energy harvesting systems. CAD methodologies have also been adopted to optimize different metrics such as schedulability, timing (efficiency), reliability, energy, and security for CPS. For instance, Wei et al. [31] propose a novel model predictive control-based algorithm for decreasing the peak energy demand and total energy consumption. Furthermore, the authors develop a system-level approach to co-schedule the usage of grid electricity and battery storage with building heating, ventilating, and HVAC systems [32]. Cao et al. [33, 34] put forward energy-efficient algorithms to schedule approximate computation tasks. As far as economic interest is concerned, the customers' home appliances are scheduled at the community level to minimize the energy purchasing expense from utilities at the market level [35].

For improving system security, Lin et al. [36] propose a general security-aware design method to enhance security with certain design constraints in a whole framework. This method can be further applied to handle a security-aware design problem for vehicle-to-vehicle communications with dedicated short-range communication technology. In order to handle pricing cyberattacks, Liu et al. [37] develop a partially observable Markov decision process-based detection algorithm with a policy transfer graph and reward expectation for the potential future impact. In addition, the authors also devise a detection technique based on partially observable Markov decision process and Bollinger bands to analyze the energy theft cyberattack [30]. Zheng et al. [38] combine control-theoretic methods at the functional layer and cybersecurity techniques at the embedded platform layer, and address security together with other design metrics such as control performance under resource and real-time constraints.

Similar to integrated circuit design, the whole process of CPS is very complex such that manual design cannot achieve the optimal solution in homogeneous domains. Given this, Sztipanovits et al. [39] build a comprehensive design tool suite for CPS. Nuzzo et al. [40] conduct a top-down mapping of application-level constraints with a bottom-up propagation of platform constraints to find the right composition of platform components that meets an application's requirements. Maasoumy et al. [41] propose a co-design approach to analyze the tight interaction between the embedded platform and the control algorithm. Deng et al. [42] propose an integrated synthesis flow to address the supply chain problem in CPS.

1.4.2 Data Analytics

Recently, machine learning technologies are developed to handle frontier challenges arising from the presence of uncertainty, multiple time-scales, multiple energy resources, the security issues, and the reliability issues related to modernized CPS. It is expected that these machine learning can improve CPS performance in terms of above aspects. To be specific, Chen et al. [43, 44] develop a linear regression model to determine the temperature variation of each location and the most important factors contributing to it. Lin et al. [45] adopt an adaptive model to reduce the huge computational cost. A classification is proposed by Sridhar et al. [46] to identify cyberattack by highlighting dependencies within the cyber-physical control parts required to facilitate the smart grid computations and communication. A supervised learning algorithm is developed by Jones et al. [47] to infer formulae to distinguish between desirable and undesirable behavior. Zhang et al. [29] utilize a generic model to identify electric theft under imbalance data. Esmalifalak et al. [48] adopt both supervised and unsupervised machine learning techniques for stealthy attack detection.

1.4.3 Hardware Support

In CPS, many diverse cyber devices are required to meet the increasing QoS requirements. As mentioned in the previous section, CPS includes three kinds of devices, i.e., sensors, communications systems, and processors. With the rapid development of semiconductor fields, sensors become more and more sensitive, precise, adaptive, and intelligent. For example, wearable equipment like smart watch have many emerging functions such as monitoring sphygmus, blood pressure, and sleep time. The data collected by these functions can be easily sent to a computing center for medical purposes, with the help of real-time efficient communication systems.

The limited computational resource also brings challenges in CPS. Specifically, with the development of machine learning and deep learning, the collected data can be accurately analyzed. However, the hardware platform is very expensive and energy-consuming. In recent years, some researchers (e.g., Zhang et al. [49]) seek to deploy deep learning algorithms onto computational resources limited hardware platforms, which provides many opportunities in broadening the application domains of CPS for high performance computing and data analysis.

1.5 Conclusion

In this chapter, we give a high level review of CPS development and research directions from the aspects of applications, QoS, and methodologies. We first summarize recent works on classical CPS applications of smart grid, smart building, electric vehicle, as well as health and medicine. We then discuss the works on improving reliability and security of CPS. We finally discuss CPS design methodologies from the perspectives of design automation, data analysis, and hardware support.

References

1. Q. Zhu, A. Sangiovanni-Vincentelli, S. Hu, X. Li, Design automation for cyber-physical systems. Proc. IEEE **106**(9), 1479–1483 (2018)
2. V.C. Gungor, D. Sahin, T. Kocak, S. Ergut, C. Buccella, C. Cecati, G.P. Hancke, Smart grid technologies: communication technologies and standards. IEEE Trans. Ind. Inf. (TII) **7**(4), 529–539 (2011)
3. Y. Liu, S. Hu, H. Huang, R. Ranjan, A. Zomaya, L. Wang, Game theoretic market driven smart home scheduling considering energy balancing. IEEE Syst. J. (ISJ) **11**(2), 910–921 (2017)
4. N. Komninos, E. Philippou, A. Pitsillides, Survey in smart grid and smart home security: issues, challenges and countermeasures. IEEE Commun. Surv. Tutorials (CMOST) **16**(4), 1933–1954 (2014)
5. P. Cong, L. Li, J. Zhou, K. Cao, T. Wei, M. Chen, S. Hu, Developing user perceived value based pricing models for cloud markets. IEEE Trans. Parallel Distrib. Syst. (TPDS) **29**(4), 2742–2756 (2018)
6. T. Wang, J. Zhou, G. Zhang, T. Wei, S. Hu, Customer perceived value- and risk-aware multiserver configuration for profit maximization. IEEE Trans. Parallel Distrib. Syst. (TPDS) **31**(5), 1074–1088 (2020)

7. B. Zheng, P. Deng, R. Anguluri, Q. Zhu, F. Pasqualetti, Cross-layer codesign for secure cyber-physical systems. IEEE Trans. Comput. Aided Des. Integr. Circuits Syst. (TCAD) **35**(5), 699–711 (2016)
8. J. Kleissl, Y. Agarwal, Cyber-physical energy systems: focus on smart buildings, in *Proceedings of ACM/IEEE Design Automation Conference (DAC)* (ACM, New York, 2010), pp. 749–754
9. X. Chen, X. Li, S. X.-D. Tan, From robust chip to smart building: CAD algorithms and methodologies for uncertainty analysis of building performance, in *Proceedings of IEEE/ACM International Conference on Computer-Aided Design (ICCAD)*, pp. 457–464 (2015)
10. B. Lin, B. Yu, Smart building uncertainty analysis via adaptive lasso. IET Cyber-Phys. Syst. Theory Appl. **2**(1), 42–48 (2017)
11. T. Wei, X. Chen, X. Li, Q. Zhu, Model-based and data-driven approaches for building automation and control, in *Proceedings of IEEE/ACM International Conference on Computer-Aided Design (ICCAD)* (ACM, New York, 2018), p. 26
12. M.B. Alawieh, X. Tang, D.Z. Pan, S 2-PM: semi-supervised learning for efficient performance modeling of analog and mixed signal circuits, in *Proceedings of IEEE/ACM Asia and South Pacific Design Automation Conference (ASPDAC)* (ACM, New York, 2019), pp. 268–273
13. W. Shan, X. Liu, M. Lu, L. Wan, J. Yang, A low-overhead timing monitoring technique for variation-tolerant near-threshold digital integrated circuits. IEEE Access **6**, 138–145 (2018)
14. R. Rajkumar, I. Lee, L. Sha, J. Stankovic, Cyber-physical systems: the next computing revolution, in *Proceedings of ACM/IEEE Design Automation Conference (DAC)* (IEEE, Piscataway, 2010), pp. 731–736
15. A. Baronchelli, R. Ferrer-i Cancho, R. Pastor-Satorras, N. Chater, M.H. Christiansen, Networks in cognitive science. Trends Cogn. Sci. **17**(7), 348–360 (2013)
16. R. Peeling, D. Mabey, Point-of-care tests for diagnosing infections in the developing world. Clin. Microbiol. Infect. **16**(8), 1062–1069 (2010)
17. Y. Luo, K. Chakrabarty, T.-Y. Ho, Dictionary-based error recovery in cyberphysical digital-microfluidic biochips, in *Proceedings of IEEE/ACM International Conference on Computer-Aided Design (ICCAD)* (ACM, New York, 2012), pp. 369–376
18. C.H. Ahn, J.-W. Choi, G. Beaucage, J.H. Nevin, J.-B. Lee, A. Puntambekar, J.Y. Lee, Disposable smart lab on a chip for point-of-care clinical diagnostics. Proc. IEEE **92**(1), 154–173 (2004)
19. L. Wu, G. Kaiser, An autonomic reliability improvement system for cyber-physical systems, in *Proceedings of IEEE International Symposium on High-Assurance Systems Engineering (HASE)*, pp. 56–61 (2012)
20. K. Cao, G. Xu, J. Zhou, M. Chen, T. Wei, K. Li, Lifetime-aware real-time task scheduling on fault-tolerant mixed-criticality embedded systems. Futur. Gener. Comput. Syst. (FGCS) **100**, 165–175 (2019)
21. K. Cao, J. Zhou, P. Cong, L. Li, T. Wei, M. Chen, S. Hu, X. Hu, Affinity-driven modeling and scheduling for makespan optimization in heterogeneous multiprocessor systems. IEEE Trans. Comput. Aided Des. Integr. Circuits Syst. (TCAD) **38**(7), 1189–1202 (2019)
22. K. Cao, J. Zhou, T. Wei, M. Chen, S. Hu, K. Li, A survey of optimization techniques for thermal-aware 3D processors. J. Syst. Archit. (JSA) **97**, 397–415 (2019)
23. J. Zhou, K. Cao, P. Cong, T. Wei, M. Chen, G. Zhang, J. Yan, Y. Ma, Reliability and temperature constrained task scheduling for makespan minimization on heterogeneous multi-core platforms. J. Syst. Softw. (JSA) **133**, 1–16 (2017)
24. J. Zhou, K. Cao, J. Sun, Y. Zhang, T. Wei, A framework to solve the energy, makespan and lifetime problems in reliability-driven task scheduling, in *Proceedings of IEEE Cyber, Physical and Social Computing (CPSCom)* (IEEE, Piscataway, 2019), pp. 608–614
25. M. Haque, H. Aydin, D. Zhu, On reliability management of energy-aware real-time systems through task replication. IEEE Trans. Parallel Distrib. Syst. (TPDS) **28**(3), 813–825 (2017)

26. L. Li, P. Cong, K. Cao, J. Zhou, T. Wei, M. Chen, Feedback control of real-time EtherCAT networks for reliability enhancement in CPS, in *Proceedings of IEEE/ACM Design, Automation and Test in Europe Conference and Exhibition (DATE)* (IEEE, Piscataway, 2018), pp. 688–693

27. L. Li, P. Cong, K. Cao, J. Zhou, T. Wei, M. Chen, S. Hu, X. Hu, Game theoretic feedback control for reliability enhancement of EtherCAT-based networked systems. IEEE Trans. Comput. Aided Des. Integr. Circuits Syst. (TCAD) **38**(9), 1599–1610 (2019)

28. Y. Liu, Y. Zhou, S. Hu, Combating coordinated pricing cyberattack and energy theft in smart home cyber-physical systems. IEEE Trans. Comput. Aided Des. Integr. Circuits Syst. (TCAD) **37**(3), 573–586 (2018)

29. Q. Zhang, M. Zhang, T. Chen, J. Fan, Z. Yang, G. Li, Electricity theft detection using generative models, in *IEEE International Conference on Tools with Artificial Intelligence (ICTAI)* (IEEE, Piscataway, 2018), pp. 270–274

30. Y. Liu, S. Hu, Cyberthreat analysis and detection for energy theft in social networking of smart homes. IEEE Trans. Comput. Soc. Syst. (TCSS) **2**(4), 148–158 (2015)

31. T. Wei, Z. Qi, M. Maasoumy, Co-scheduling of HVAC control, EV charging and battery usage for building energy efficiency, in *Proceedings of IEEE/ACM International Conference on Computer-Aided Design (ICCAD)* (IEEE, Piscataway, 2014), pp. 191–196

32. T. Wei, T. Kim, S. Park, Z. Qi, X.D. Tan, N. Chang, S. Ula, M. Maasoumy, Battery management and application for energy-efficient buildings, in *Proceedings of ACM/IEEE Design Automation Conference (DAC)* (ACM, New York, 2014), pp. 1–6

33. K. Cao, G. Xu, J. Zhou, T. Wei, M. Chen, S. Hu, QoS-adaptive approximate real-time computation for mobility-aware IoT lifetime optimization. IEEE Trans. Comput. Aided Des. Integr. Circuits Syst. (TCAD) **38**(10), 1799–1810 (2019)

34. K. Cao, J. Zhou, G. Xu, T. Wei, S. Hu, Exploring renewable-adaptive computation offloading for hierarchical QoS optimization in FOG computing. IEEE Trans. Comput. Aided Des. Integr. Circuits Syst. (TCAD) (2019). https://doi.org/10.1109/TCAD.2019.2957374

35. Y. Liu, S. Hu, H. Huang, R. Ranjan, A.Y. Zomaya, L. Wang, Game-theoretic market-driven smart home scheduling considering energy balancing. IEEE Syst. J. (ISJ) **11**(2), 910–921 (2017)

36. C.W. Lin, B. Zheng, Q. Zhu, A. Sangiovanni-Vincentelli, Security-aware design methodology and optimization for automotive systems. ACM Trans. Des. Autom. Electron. Syst. (TODAES) **21**(1), 1–26 (2015)

37. Y. Liu, S. Hu, T.Y. Ho, Leveraging strategic detection techniques for smart home pricing cyberattacks. IEEE Trans. Dependable Secure Comput. (TDSC) **13**(2), 220–235 (2016)

38. B. Zheng, D. Peng, R. Anguluri, Z. Qi, F. Pasqualetti, Cross-layer codesign for secure cyber-physical systems. IEEE Trans. Comput. Aided Des. Integr. Circuits Syst. (TCAD) **35**(5), 699–711 (2016)

39. J. Sztipanovits, T. Bapty, S. Neema, X. Koutsoukos, E. Jackson, Design tool chain for cyber-physical systems: lessons learned, in *Proceedings of ACM/IEEE Design Automation Conference (DAC)* (ACM, New York, 2015), p. 81

40. P. Nuzzo, A.L. Sangiovanni-Vincentelli, D. Bresolin, L. Geretti, T. Villa, A platform-based design methodology with contracts and related tools for the design of cyber-physical systems. Proc. IEEE **103**(11), 2104–2132 (2015)

41. M. Maasoumy, Z. Qi, L. Cheng, F. Meggers, A. Vincentelli, Co-design of control algorithm and embedded platform for building HVAC systems, in *Proceedings of International Conference on Cyber-Physical Systems (ICCPS)* (ACM, New York, 2013), pp. 61–70

42. P. Deng, F. Cremona, Q. Zhu, M. Di Natale, H. Zeng, A model-based synthesis flow for automotive CPS, in *Proceedings of International Conference on Cyber-Physical Systems (ICCPS)* (ACM, New York, 2015), pp. 198–207

43. X. Chen, X. Li, S.X.-D. Tan, Overview of cyber-physical temperature estimation in smart buildings: from modeling to measurements, in *Proceedings of IEEE Conference on Computer Communications Workshops (INFOCOM WKSHPS)* (IEEE, Piscataway, 2016), pp. 251–256

44. X. Chen, X. Li, S.X.-D. Tan, From robust chip to smart building: CAD algorithms and methodologies for uncertainty analysis of building performance, in *Proceedings of IEEE/ACM International Conference on Computer-Aided Design (ICCAD)* (IEEE Press, Piscataway, 2015), pp. 457–464

45. B. Lin, B. Yu, Smart building uncertainty analysis via adaptive lasso. IET Cyber-Phys. Syst. Theory Appl. **2**(1), 42–48 (2017)

46. S. Sridhar, A. Hahn, M. Govindarasu, Cyber–physical system security for the electric power grid. Proc. IEEE **100**(1), 210–224 (2012)

47. A. Jones, Z. Kong, C. Belta, Anomaly detection in cyber-physical systems: a formal methods approach, in *Proceedings of IEEE Conference on Decision and Control (CDC)* (IEEE, Piscataway, 2014), pp. 848–853

48. M. Esmalifalak, L. Liu, N. Nguyen, R. Zheng, Z. Han, Detecting stealthy false data injection using machine learning in smart grid. IEEE Syst. J. (ISJ) **11**(3), 1644–1652 (2017)

49. Q. Zhang, M. Zhang, T. Chen, Z. Sun, Y. Ma, B. Yu, Recent advances in convolutional neural network acceleration. Neurocomputing **323**, 37–51 (2019)

Chapter 2
A Utility-Driven Data Transmission Optimization Strategy in Large-Scale Cyber-Physical Systems

Soumi Chattopadhyay, Ansuman Banerjee, Nilanjan Banerjee, and Bei Yu

2.1 Introduction

Large-scale sensor-enabled context-aware CPS [1] is fast emerging as the preferred vehicle of operation in the era of Internet-of-Things. Digitally instrumented devices are being built into the fabric of our environments like never before and used to continuously monitor, manage, and regulate physical flows and processes, often in real-time. Common examples of such real-time contextual intelligence in action are smart traffic management, smart water systems, smart health-care, smart rooms and offices and, more recently, smart cities [2].

In this chapter, we deal with a simple operational model of a CPS that comprises of a set of distributed sensors communicating with a back-end controller to monitor and regulate system parameters based on some system objectives. The operations of these systems involve sensing, networking, and processing of significant amount of data. The inputs are received from a set of information observers, often called the front-end (e.g., sensors, monitors, data sources, network feeds, etc.). The back-end is usually a processing unit (running either on cloud servers or on mobile gateway/edge devices [3]) that accumulates all observations received over the network from

S. Chattopadhyay (✉)
Indian Institute of Information Technology Guwahati, India

A. Banerjee
ACMU, Indian Statistical Institute, Kolkata, India

N. Banerjee
Onometra Technologies Pvt. Ltd., India

B. Yu
Department of Computer Science and Engineering, The Chinese University of Hong Kong, Shatin, Hong Kong
e-mail: byu@cse.cuhk.edu.hk

© Springer Nature Switzerland AG 2020
S. Hu, B. Yu (eds.), *Big Data Analytics for Cyber-Physical Systems*,
https://doi.org/10.1007/978-3-030-43494-6_2

13

different disparate sources, and monitors/evaluates a set of system objectives for deciding on necessary actuations. Examples [4] of such systems are ubiquitous from large-scale weather monitoring systems, fleets of vehicles, to a connected automotive and a sensor-fitted human body network. IBM and the city government of Rio de Janeiro, Brazil formed a partnership to create a city wide instrumented system that draws together data streams from 30 agencies including traffic and public transport, municipal and utility services, emergency services, weather feeds, information sent in by employees and public via phone, internet, and radio into a single data processing and analytics center.[1] A similar effort led by Intel aims at using London as a testbed where users and the existing city infrastructure are exploited to improve the efficiency of urban processes. Needless to say that smart city platforms are going to be further enriched with more number of data sources providing richer information about the cities and the citizens.

With such extensive instrumentation of the world around us, the volume and quality of data collected has grown exponentially. Management and interpretation of data of such magnitudes in real-time is a stupendous challenge today in the CPS context. While computational scalability has received most of the recent research attention in the effort of designing better CPS, communication scalability of information flow processing has also been looked at [5–7]. Indeed, with the pressing need to extend the lifetime of the sensors (these are typically of low power), a significant body of research [8, 9] has focused on efficient means of regulating message observation and transmission using energy reducing optimizations, efficient query processing, or classical compression techniques. Figure 2.1 shows an architectural overview of such systems. In summary, it performs three major tasks, namely (a) capturing of data from the environment, (b) transmission of data over the network for processing, and (c) back-end processing of the transmitted data for necessary actions.

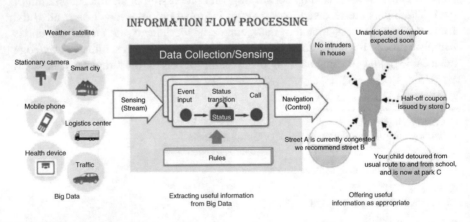

Fig. 2.1 System architecture

In this work, we propose a fundamentally different operational model for data dissemination optimization for a large-scale distributed CPS. Our main idea here is to identify the relevance of each sensor observation as far as the objectives in the back-end are concerned. By transmitting only the necessary set of events and thereby avoiding redundancies makes the back-end computation more efficient. The motivation behind this work stems from the observation that not all data generated in the system is always important, as far as the end objectives being monitored at the back-end are concerned, and hence, need not be distributed always. This saves network resources. The basic tenet of our system is to associate with each event collected at a source, a notion of spatial and temporal relevance in real-time, and measure its effective utility, which can then lead to significant reductions of data transmission as evident from the experimental results as well. We demonstrate the effectiveness of our proposal on two real-life case studies. The main contributions of this work are as follows:

- A formal system model based on mathematical logic and its transformations for modeling system objectives for a CPS system.
- Using formal methods for computing minimal subsets of events to be observed, based on the system objectives.
- Using logic transformations for regulating information capture and flow.
- Reducing the number of computations at the back-end, based on content of information generated.
- A predictive model for CPS that builds on event history and patterns and attempts to predict the events that are required for current evaluation.

To the best of our knowledge, only a few articles have looked at the problem from a similar perspective as ours, with the motivation of analyzing the system objectives and deciding on data dissemination. Most of the proposals in this direction have a topic-based filtering approach [10, 11] that reduce the overall communication overhead in the network. In our first work [12], we have proposed an early foundation of a static analysis and optimization strategy for message transmission in a context-aware system. The second work [13] includes a more detailed discussion and formal proofs of the proposed strategy [12] along with additional optimization techniques.

This chapter is organized as follows. Section 2.2 presents a motivating example. Section 2.3 presents the formal system model. Section 2.4 presents the different optimization techniques proposed in this work, while Sect. 2.5 presents two detailed case studies on our CPS proposal. Section 2.6 presents an overview of the implementation and results.

2.2   A Motivating Example

We explain the motivation for designing an efficient information flow processing engine on a simple hypothetical example of an intelligent system for a smart home. Consider a scenario with five devices, namely an air-conditioner (AC), a fan, a light, a curtain, and a camera device, communicating with a central home automation controller. We assume that the sensors attached to the devices (the information observers) act as the information observers and transmitters, and are designated to monitor and send observations on events in the room, while the controller is the processing back-end, which contains the information processing engine. We assume for the sake of simplicity that transmissions are through a reliable network and there are no message losses. We now describe the responsibility of each device. The AC can sense and communicate the temperature and humidity of the room. The fan and the light send information on their current state, i.e., whether they are on or off. The curtain can communicate its state as well, i.e., whether it is open or closed. Additionally, it can also sense and communicate whether the outside temperature is greater than the inside temperature. The camera device can sense and communicate if there is any person in the room. The back-end information processing engine inside the automation controller contains a set of rules and the corresponding actuations which are basically *If-then* rules on the information observation data. If the rule evaluates to true, necessary action is triggered.

- *Rule 1*: Turn on AC if the current state of the AC is off and any of the following hold:

  - Room temperature $\geq 30\,°C$ and humidity $\geq 80\%$.
  - Room temperature $\geq 30\,°C$, the curtain is open and outside temperature > inside temperature.
  - Room temperature $\geq 30\,°C$, humidity $\geq 80\%$, the curtain is currently closed and the fan is off.

- *Rule 2*: Turn off the light if:

  - The room is empty and the light is currently on.

The example presented here may appear to be non-intuitive to the reader, but, in reality, a general system designer is equipped with very limited exposure to sophisticated logic and specification formalisms to be able to write rules that are easily amenable to automated analysis. The example presented here is representative of the formal training (or lack of it) available to the designer. The rules are usually expressed by domain experts (and not computer scientists/engineers/logicians) who model the system in natural language forms. We show here how a formal foundation can lead to remarkable efficiency in system performance as far as the transmission profile is concerned.

We now analyze the situation from the perspective of the devices, which are entrusted with the responsibility of observing the parameters and communicating the values to the central controller, whenever there is a change in the value of the parameters they observe/communicate. Values received from the devices are used by the information flow processing engine to evaluate the two rules and accordingly decide the next course of action. We define a few Boolean predicates in Table 2.1 to model some conditions on the variables observed by the devices. These predicates are used in the rules. The devices evaluate the values (*true/false*) of the predicates,

Table 2.1 The Boolean predicates

p_1: Room temperature $\geq 30\,°C$
p_2: Humidity $\geq 80\%$
p_3: The curtain is open
p_4: Outside temperature $>$ inside temperature
p_5: Fan is on
p_6: AC is on
p_7: The room is empty
p_8: Light is on

Table 2.2 Random snapshot of ten time instants

Time instant	p_1	p_2	p_3	p_4	p_5	p_6	p_7	p_8
1	1	1	1	0	1	0	1	1
2	1	1	0	1	0	1	1	0
3	0	1	0	0	0	1	1	0
4	0	1	1	1	1	1	1	0
5	0	0	0	0	0	1	1	0
6	0	1	1	1	0	1	1	1
7	0	1	1	0	1	1	1	1
8	0	0	0	1	1	1	1	1
9	0	0	1	0	0	1	1	0
10	0	0	1	1	0	1	1	0
Total changes	2	4	7	10	6	2	1	4

before sending the update to the back-end controller. On receiving these updates, the controller (in particular, the back-end information processing engine) acts accordingly by evaluating the rules which are affected by these updates. The communication is done using the Boolean predicates, which takes fewer message bits, as opposed to transmitting the exact value of the observations. This could have been done using some encoding scheme for the predicates directly as well. However, we use the Boolean encoding just for the sake of simplicity of illustration.

Rewriting the rules in terms of the predicates, we have

- *Rule 1*: $(p_1.p_2 + p_1.p_3.p_4 + p_1.p_2.\bar{p}_3.\bar{p}_5).\bar{p}_6$: Turn on AC
- *Rule 2*: $p_7 + p_8$: Turn off light.

Here, . represents the AND operator and + represents the OR operator. For the purpose of illustration of the number of transmissions, we consider random values on the predicates. As mentioned previously, the predicates are derived from the observations at the devices. Table 2.2 contains a randomly generated snapshot of ten time instants. Each row in the table shows the values assigned to the eight predicates that we define in Table 2.1. Each entry in the table contains a 0 or 1 depending on whether the corresponding Boolean predicate is true or false. For example, if Room temperature $\geq 30\,°C$, the predicate p_1 is 1, otherwise p_1 is 0.

2.2.1   How Do Standard Transmission Methods Work?

To motivate our proposal, we first explain how standard transmission procedures work for the predicate profile shown in Table 2.2. In a synchronous round-based transmission scheme, in each round, the values of all the predicates are transmitted, even if none of them change. A large number of transmissions take place in this case. In the case of a change-driven partially synchronous round-based transmission scheme, whenever there is a change in the value of any predicate, a transmission message is triggered, with the message carrying the latest valuations of all the predicates (not only the values of the predicates which have changed, but all of them). Quite evidently, this is not quite efficient as well. In an asynchronous transmission scheme, whenever any predicate changes from its current value, only the value of the changed predicate (and not the entire snapshot) is transmitted. Consider the snapshot shown in Table 2.2. On one extreme, in the completely synchronous round-based transmission scheme, a total of 80 (in each of the 10 slots, all the 8 predicates are transmitted) transmissions are required. For an asynchronous transmission scheme, 36 transmissions are required as shown in Table 2.2.

2.2.2   Our Method at Work

Our method is motivated by the fact that not all predicates need to be observed/transmitted or processed at all times, even if they have changed. If we analyze the rules, we can reduce the number of transmissions. If we observe rule 1, we see that if the third condition of the first rule holds, i.e., Room temperature $\geq 30\,^\circ$C, humidity $\geq 80\%$, the curtain is closed and the fan is off, then the first condition is automatically guaranteed. Conversely, we can say if any of the elements in the first condition of rule 1 is false, the third condition can never be true. Hence, to evaluate rule 1, the first two conditions are sufficient and the third condition is redundant. From the above observation, we can conclude that the state of the fan is redundant for the information flow processing engine for evaluation of rule 1. As we show later, if a predicate is redundant for all the rules at the back-end, we need not observe that predicate at all.

While the above observations can be arrived at by a static analysis/redundancy check on the given set of rules, we also show how it is possible to reduce the number of transmissions further by analyzing the devices' observations at the current snapshot. We can see that the curtain communicates its state and whether the outside temperature > room temperature. If any of the predicates p_3 and p_4 become false, the information flow processing engine will never perform an actuation due to the second condition of rule 1. Hence, instead of sending the individual status of these predicates, the curtain can send false if either of them is false and true if both of them are true, since both of these are observed and transmitted by the curtain itself.

The information flow processing engine does not need to know exactly which of the predicates is true. This can save message bits as well.

While the above discussion is about reducing the transmission overhead based on the rule structure, it is also possible to reduce the number of transmissions further, based on the semantics of the rules and values of some predicates, as we discuss in the following. In each round, the information processing engine receives the status of the transmitted predicates and evaluates the rules, assuming the values of the other predicates (not received) are same as earlier. We go ahead one step and comment on how the information engine can cut down on the transmissions by selectively directing the information observers to abstain from sending some predicates, even if they have changed. Consider rule 1 as an example. It can be observed that if the AC is on, then the first rule can never be true. Once the information that AC is on is received, the information flow processing engine does not need any other observation about room temperature, humidity, or the curtain state in order to evaluate the first rule. If we observe the second rule, we see that two predicates are required to decide whether to turn off the light. However, as long as the room is non-empty, the information flow processing engine can evaluate this rule without knowing the status of the light.

We formalize the algorithm for selective transmission in this chapter based on the above observations. Specifically, to capture the relevance of a predicate, we define the notion of the *effective utility* of an information observation and subsequent transmission, based on the relevance of transmission of the predicates appearing in the rules. With our proposed optimizations, 19 transmissions are required, as shown in Table 2.3, where each cell entry denotes the value transmitted to the back-end (X indicates no transmission). In this work, we put forward a formal framework for reducing the number of transmissions, as illustrated in this example. The information flow processing engine proposed in this work needs to capture only seven predicates over the period of ten time instants for this snapshot, as described in Table 2.4.

Table 2.3 Transmission of predicates in our method

Time instant	p_1	p_2	$p_3.p_4$	p_5	p_6	p_7	p_8
1	1	1	0	X	0	1	1
2	X	X	X	X	1	X	0
3	0	X	X	X	X	X	X
4	X	X	1	X	X	X	X
5	X	0	0	X	X	X	X
6	X	1	1	X	X	X	1
7	X	X	0	X	X	X	X
8	X	0	X	X	X	X	X
9	X	X	X	X	X	X	0
10	X	X	1	X	X	X	X
Total transmissions	2	4	6	0	2	1	4

Table 2.4 Processing profile of the information flow processing engine in our method

Time instant	p_1	p_2	$p_3 . p_4$	p_5	p_6	p_7	p_8
1	1	1	0	X	0	1	1
2	X	X	X	X	1	X	X
3	X	X	X	X	X	X	X
4	X	X	X	X	X	X	X
5	X	X	X	X	X	X	X
6	X	X	X	X	X	X	X
7	X	X	X	X	X	X	X
8	X	X	X	X	X	X	X
9	X	X	X	X	X	X	X
10	X	X	X	X	X	X	X
Total transmissions	1	1	1	0	2	1	1

2.3   Formal Model

We consider the following system workflow of an information flow processing engine in this work.

- A set of information observers collect observations about a set of parameters. Based on the values observed, they evaluate and transmit the truth values of a set of predicates. The observations may be of arbitrary data type but the predicates are Boolean.
- The information observers communicate with a back-end processing engine through a reliable network.
- The back-end has a set of rules $\hat{C}$ and corresponding actions. The rules are expected to be exercised in specific contexts as expressed in their triggering conditions.
- Based on the values of the predicates, the back-end engine evaluates the rules and based on the evaluation status of the rules, triggers necessary actuations.

At the back-end, before the network is initialized, the processing engine pre-processes the rules using our proposed methods and communicates to the infor-mation observers, which predicates (and hence, which observations) are needed and at what instants. The overall objective of our scheme is to save redundant transmissions. Based on the values received from the information observers, the processing engine evaluates the rules and takes necessary steps. At the observer end, the necessary observations at the appropriate instants (as communicated to them by the back-end) are made, and transmitted if needed. We illustrate our proposal in the following sections. We begin by defining formally the syntax and semantics of a data predicate, a rule, and an action on a set of data variables. A *data variable* is a parameter observed by some information observer. Examples of data variables include temperature, humidity, CO_2 level, etc.

Definition 1 (Data Predicate) A data predicate is either itself *true*, *false*, or an expression of the form $< \text{lexp} > \bowtie < \text{rexp} >$, where $\bowtie \in \{=, \neq, <=, >=, <, >\}$, $< \text{lexp} >$ and $< \text{rexp} >$ are data variables or constants. ∎

Given a valuation to its variables, a data predicate evaluates to *true* or *false*. The truth value of these predicates is transmitted to the back-end processing engine.

Example 1 Consider a variable u which denotes the temperature of the system. A data predicate $\mathcal{P}$ is defined as: $\mathcal{P} : u < 20\,°\text{C}$. If the temperature is less than $20\,°\text{C}$ at any instant, $\mathcal{P}$ is evaluated as *true*, else it is *false*. ∎

Definition 2 (Rule) A rule is a Boolean combination of data predicates. ∎

Definition 3 (Rule-Triggered Action) A rule-triggered action is an expression of the form *Rule: Action*. ∎

Action is the resulting actuation performed by the back-end in response to a rule. The intuitive semantics of a rule-triggered action is as follows: the *Action* will be executed only if the predicate values at the current snapshot make the *rule* evaluate to *true*. We do not consider any specific form of the *Action*. For example, it can be a signal generated at the back-end or some tasks initiated, when the rule evaluates to *true*.

Example 2 An example rule-triggered action follows:

Rule: Room temperature $>20\,°\text{C}$ and AC is off.
Action: Turn on AC. ∎

2.4 The Proposed Methodology

We explain below the various optimization techniques we have used to form our strategy for reducing message transmissions proposed in this chapter. The intuition behind the optimization techniques is to analyze the back-end rule repository and derive the effective importance of the predicates (in other words, the observed parameters) with respect to the actuation of the rule actions. As a result of this analysis, we obtain a much reduced transmission profile as compared to other schemes. We begin the discussion of our proposed scheme with a few definitions, which help us build the foundations of our analysis framework. Let $\mathcal{P}$ be the set of data predicates and $\hat{\mathcal{C}}$ be the set of rules.

Definition 4 (Co-factor) The positive (negative) co-factor of a rule $\mathcal{C}$ defined over a set of Boolean predicates $\mathcal{P} = \{p_1, p_2, \ldots p_n\}$ with respect to a data predicate $p_i \in \mathcal{P}$ is obtained by substituting 1 (true) or 0 (false) in $\mathcal{C}$. ∎

The positive co-factor, denoted by $\mathcal{C}_{p_i}$, is obtained as $\mathcal{C}_{p_i} = \mathcal{C}(p_1, p_2, \ldots, p_i = 1, \ldots, p_n)$. Similarly, the negative co-factor, denoted as $\mathcal{C}_{\bar{p}_i}$, is obtained as $\mathcal{C}(p_1, p_2, \ldots, p_i = 0, \ldots, p_n)$. The co-factors are independent of the predicate

p_i with respect to which they are computed. We now define the concept of decomposition of a rule, similar to Shannon's expansion of Boolean functions [14].

Definition 5 (Rule Decomposition) The decomposition of a rule C with respect to a predicate $p \in \mathcal{P}$ is obtained as: $C = p.C_p + \bar{p}.C_{\bar{p}}$, where C_p and $C_{\bar{p}}$ are, respectively, the positive and negative co-factors of C with respect to p. ∎

Example 3 Consider the rule $C = p_5 + p_6 + p_7.p_8$. The positive co-factor of C with respect to p_5 is $1 + p_6 + p_7.p_8 = 1$. The negative co-factor is $0 + p_6 + p_7.p_8$ $= p_6 + p_7.p_8$. The decomposition of C with respect to p_5 is $C = p_5.C_{p_5} + \bar{p}_5.C_{\bar{p}_5}$ $= p_5.1 + \bar{p}_5.(p_6 + p_7.p_8) = p_5 + p_6 + p_7.p_8$. ∎

2.4.1 Optimizations Based on Rule Co-factor Analysis

We begin our proposal of reducing message transmissions with a simple observation in terms of co-factors. Consider the current value of $p = 1$ for some predicate $p \in \mathcal{P}$ and also consider for some rule C, C_p is a function of $p_{i_1}, \ldots, p_{i_l}$. In this case, to evaluate C, the back-end does not need to know the status of the predicates in $\mathcal{P} \setminus \{\{p_{i_1}, \ldots, p_{i_l}\} \bigcup \{p\}\}$ until p changes, where $\setminus$ is the set difference operation. If this information is made available to the information observers, it may result in quite a significant reduction in the number of transmissions, and this is what we use in our method. If some information observer can observe all these remaining set of predicates $\mathcal{P} \setminus \{\{p_{i_1}, \ldots, p_{i_l}\} \bigcup \{p\}\}$, it can judiciously decide when to send observations next, which in turn leads to reduction in the number of transmissions through the network. This can be done in the pre-processing step itself when the network is initialized, and kept as a look-up table at the information observer end or at run-time when the rule evaluates to 1. If these set of predicates are not co-observed, in other words, multiple information observers capture their values, the back-end can send this information to the corresponding observers at run-time, mentioning that they will be polled when needed. As long as the back-end does not receive a changed value of p, it does not poll these other information observers, and hence, in turn, observations and message transmissions are both reduced. For example, when C_p or $C_{\bar{p}}$ equals to a constant, i.e., 0 or 1, this means the change of value of other predicates is not useful for C as long as the value of p remains same. So to evaluate C, the back-end does not need the status of any other predicate other than p until it changes. Hence, the information observer observing this predicate p can keep on observing p, but transmit only when it changes. At the initialization stage, the back-end analyzes the rules and gathers the set of predicates whose positive or negative co-factors evaluate to a constant. This information is passed on to the information observers, and each observer precisely knows when to transmit the valuations of the other predicates it observes. Depending on the current snapshot, the back-end informs the information observers about the required set of predicates.

Optimization Observation 1 Let C be a rule and p be a data predicate. Based on the value of $p = 1$ (or 0), let C_p (or $C_{\bar{p}}$) be a function of $p_{i_1}, \ldots, p_{i_l}$. To evaluate C, the back-end does not require to know the status of the predicates in $\mathcal{P} \setminus \{\{p_{i_1}, \ldots, p_{i_l}\} \bigcup \{p\}\}$ until p changes. ∎

Example 4 Consider rule 2 in Sect. 2.3, defined over the predicates p_5, p_6, p_7, p_8 as: $C(p_5, p_6, p_7, p_8) = p_5 + p_6 + p_7.p_8$ The positive co-factor of C w.r.t. p_5, $C_{p_5} = 1 + p_6 + p_7.p_8 = 1$, negative co-factor of C w.r.t. p_5, $C_{\bar{p}_5} = 0 + p_6 + p_7.p_8 = p_6 + p_7.p_8$. So, when $p_5 = 1$, the back-end does not require to know the status of p_6, p_7, p_8 until p_5 becomes 0. The same happens for p_6: as long as $p_6 = 1$, the back-end does not need any information about p_5, p_7, p_8. ∎

The concept of rule decomposition can be extended to multiple predicates as well. The decomposition of a rule C with respect to two predicates $p, q \in \mathcal{P}$ is obtained as: $C = pq.C_{pq} + p\bar{q}C_{p\bar{q}} + \bar{p}q.C_{\bar{p}q} + \bar{p}\bar{q}C_{\bar{p}\bar{q}}$, where, $C_{pq}, C_{p\bar{q}}, C_{\bar{p}q}, C_{\bar{p}\bar{q}}$ are the co-factors with respect to multiple predicates. The co-factors are obtained as follows: C_{pq} by substituting the value of $p = 1$ and $q = 1$ in C, $C_{p\bar{q}}$ by substituting the value of $p = 1$ and $q = 0$ in C, $C_{\bar{p}q}$ by substituting the value of $p = 0$ and $q = 1$, and $C_{\bar{p}\bar{q}}$ by substituting the value of $p = 0$ and $q = 0$. Co-factor analysis involving multiple predicates can help in reducing message transmissions even further. But in this case, if we want to analyze the co-factors for a rule defined over a set $\mathcal{P}$ of n predicates, we have in all an exponential number (2^n) of combinations, considering all subsets of all cardinalities of $\mathcal{P}$. In our work, we restrict ourselves to a two variable co-factor analysis. Again, as explained above, instead of looking at all the predicates in the predicate set $\mathcal{P}$ for *multiple variable co-factor analysis*, we can limit our analysis to a smaller set, containing only the predicates whose positive and negative co-factors are not constant.

Example 5 Consider rule 2 with respect to the snapshot: $p_5 = 0$, $p_6 = 0$, $p_7 = 1$, $p_8 = 1$. According to single variable co-factor analysis, $C_{\bar{p}_5}$ is a function of p_6, p_7, p_8, $C_{\bar{p}_6}$ is a function of p_5, p_7, p_8, C_{p_7} is a function of p_5, p_6, p_8, and C_{p_8} is a function of p_5, p_6, p_7. The information observers need to transmit the status of all the four predicates, since none of the co-factors are constant. *Multiple variable co-factor analysis* reveals that $C_{p_7 p_8} = p_5 + p_6 + 1.1 = 1$. So, as long as p_7 and p_8 remain 1, the back-end does not need the information of p_5 and p_6. If any of them changes, the back-end requires information of both p_5 and p_6. ∎

For each rule, we first analyze the single variable co-factors and create a decision table which contains two fields: the value of the predicate and depending on that value of the predicate, which predicates are not required to evaluate the rule. This is followed by a two variable co-factor analysis using the predicates whose positive and negative co-factors are not constant. If the set contains v number of predicates, then we have $^vC_2$ combinations and, for each combination, there exists four different values (corresponding to the four co-factors). The results of two variable co-factor analysis are stored in the decision table in the same format.

2.4.2   Optimizations Based on Unateness Analysis

We now define the concept of unateness of a rule and use it to further reduce the number of transmissions.

Definition 6 (Unate and Binate Rule) A rule C over a set of data predicates is unate if in its canonical [15] representation, every data predicate appearing in C is in exactly one polarity (either all positive or all negative). A rule that is not unate is binate.   ∎

Corresponding to any Boolean expression, we have many representations which ensure the property of canonicity, in other words, two rules which are semantically the same have identical representations in these representations. In our work, we use reduced ordered binary decision diagrams [16], which guarantees the canonicity property.

Definition 7 (Unateness) A rule C is positive unate in p_i if changing the value of p_i from 0 to 1 keeps C constant or changes C from 0 to 1.   ∎

This essentially means the following for a given rule $C(p_1, p_2, \ldots, p_{(i-1)}, p_i, p_{(i+1)}, \ldots, p_m)$: $C(p_1, \ldots, p_{(i-1)}, 1, \ldots, p_m) \geq C(p_1, \ldots, p_{(i-1)}, 0, \ldots, p_m)$. The concept of negative unateness is defined similarly, i.e., $C(p_1, \ldots, p_{(i-1)}, 0, \ldots, p_m) \geq C(p_1, \ldots, p_{(i-1)}, 1, \ldots, p_m)$.

Consider a scenario that a rule C is evaluated to 1 at some point of time according to the current snapshot. *Then the value of C does not change with the change of all those predicates whose current value is 0 and in which C is positive unate.* Intuitively, this is because, the current rule is already evaluated to *1* and it being positive unate in the predicate, cannot grow beyond its present value. In other words, these predicates are not needed to be transmitted. As earlier, the back-end decides on these conditions in the pre-processing stage and informs the information observers about its transmission requirements.

Optimization Observation 2 Let C be a rule, positive unate in the set of predicates $POS_c = \{p_{i_1}, p_{i_2}, \ldots, p_{i_l}\}$ and negative unate in the set of predicates $NEG_c = \{p_{j_1}, p_{j_2}, \ldots, p_{j_l}\}$. The transmission optimization can be done as follows:

- If C evaluates to *1* based on the current snapshot, predicates belonging to POS_c which have their current value as 0 are not required at the back-end to evaluate C.
- If C evaluates to *0* based on the current snapshot, predicates belonging to POS_c which have their current value as 1 are not required at the back-end to evaluate C.
- If C evaluates to *1* based on the current snapshot, predicates belonging to NEG_c which have their current value as 1 are not required at the back-end to evaluate C.
- If C evaluates to *0* based on the current snapshot, predicates belonging to NEG_c which have their current value as 0 are not required at the back-end to evaluate C.   ∎

Example 6 Consider a rule $C = p_1.p_2.p_3.p_4 + p_5.p_6.p_7$ and the current snapshot $p_1 = 1, p_2 = 1, p_3 = 1, p_4 = 1, p_5 = 0, p_6 = 0, p_7 = 0$. On this snapshot,

C evaluates to 1. We can see that C is positive unate in $\{p_1, \ldots, p_7\}$. As long as C remains 1, value changes of variables which have their current value as 0, i.e., $\{p_5, p_6, p_7\}$, do not affect the value of C.   ∎

At run-time, on a given snapshot, the back-end may inform the information observers to abstain from sending value changes, since these are redundant from the computation perspective of the processing engine and, again, poll the information observers, as needed. Unateness analysis can be of immense help for this analysis. The rule processing and unateness sets can be derived in the pre-processing stage, and kept at the back-end, and this can help in taking decisions at run-time in specific snapshots that are amenable to such optimization opportunities. It is further interesting to note that unateness analysis takes care of the situation where co-factor analysis with respect to more than two predicates is involved. But unateness analysis cannot take care of all situations as multiple variable co-factor analysis does. Consider the following example.

Example 7 Consider the rule $C = p_0.p_1.p_2 + p_3.p_4.p_5 + p_6.p_7.p_8.p_9$ and a snapshot $(1, 1, 1, 1, 1, 1, 1, 0, 0, 1)$. Multiple variable co-factor analysis gives us the optimal set required to evaluate C, i.e., $\{p_0, p_1, p_2\}$. The back-end does not need to know the status of $\{p_3, p_4, p_5, p_6, p_7, p_8, p_9\}$. Unateness analysis reveals that the set $\{p_7, p_8\}$ is not required by the back-end, since C is positive unate in $\{p_0, p_1, p_2, p_3, p_4, p_5, p_6, p_7, p_8, p_9\}$, the current evaluated value of C is 1, and only p_7 and p_8 are 0. However, unateness analysis cannot produce the optimal set.∎

Multiple variable co-factor analysis is computationally expensive. If we want to perform k variable co-factor analysis, then we need $^nC_k \times 2^k$ computations for each rule, where n is the number of predicates associated with the rule. In the previous example, for 3 variable co-factor analysis, 960 computations are needed and 4 variable co-factor analysis requires 3360 computations. Therefore, we restrict our proposal to 2 variable co-factor analysis. In certain cases, unateness analysis can also produce the optimal solution. For example, for the snapshot $(1, 1, 1, 0, 0, 0, 0, 0, 0, 0)$ in the previous example, unateness analysis does produce the optimal set.

2.4.3   Optimizations Based on Effective Utility

Analyzing the criticality of a predicate with respect to a rule leads to further reductions. It is important to know which of the predicates in the rule can actually influence the evaluation result of the rule. This is what we define as the effective utility of the predicate with respect to the rule. There may be some predicates whose changes are not reflected in the rule at all and there may be some predicates for whom every change is crucial for the evaluation of the rule. This is handled using *derivative analysis*.

Definition 8 (Derivative) The derivative of a rule C with respect to a predicate p_i is defined as the exclusive-or of the positive and negative co-factors of C with respect to the predicate p_i, i.e., $\partial C/\partial p_i = C_{p_i} \oplus C_{\bar{p}_i}$.   ■

Definition 9 (Effective Utility) The effective utility of a predicate $p \in \mathcal{P}$ with respect to a rule C refers to how the change of the value of p affects C.   ■

The intuitive idea is as follows. p_i can only change from 0 to 1 or 1 to 0. Consider the present value of p_i as 1. Using co-factor decomposition, we have $C = (1.C_{p_i} + 0.C_{\bar{p}_i}) = C_{p_i}$. When the value of p_i changes to 0, $C = (0.C_{p_i} + 1.C_{\bar{p}_i}) = C_{\bar{p}_i}$. C changes with change in p_i if the values of the positive and negative co-factors are different. This is expressed using $C_{p_i} \oplus C_{\bar{p}_i}$. Hence, $\partial C/\partial p_i$ denotes the change of the rule C with respect to the predicate p_i.

Derivative analysis leads to interesting observations. For a predicate p_i, we have the following:

- $C_{p_i} \oplus C_{\bar{p}_i} = 0$, implies C is *independent* of p_i. This is because $C_{p_i} = C_{\bar{p}_i}$, hence C is not affected with change in p_i. This implies the back-end does not need to receive the status of p_i at any time. If the derivative of a rule with respect to a predicate is 0, then we can simply disregard the predicate for that specific rule.
- $C_{p_i} \oplus C_{\bar{p}_i} = 1$, implies the predicate p_i is *critical* to the back-end since every change in p_i is reflected in the rule C. Hence, all updates to a critical predicate have to be transmitted to the back-end.
- $C_{p_i} \oplus C_{\bar{p}_i}$ is free from p_i. In this case, the back-end requires to know the status of p_i at least once.

Optimization Observation 3 If a rule C is independent of a predicate p, the back-end does not need to know the status of p to evaluate C.   ■

Example 8 Consider the scenario that we have a single rule: $C(p_1, p_2, p_3) = p_1.p_2.\bar{p}_3 + p_1.p_2.p_3 + \bar{p}_1.\bar{p}_2.\bar{p}_3 + \bar{p}_1.\bar{p}_2.p_3$. We can analyze the rule as follows. The derivative of C w.r.t. p_3, $\partial C/\partial p_3 = C_{p_3} \oplus C_{\bar{p}_3} = 0$. Since no change of p_3 is needed for evaluating C, the back-end does not require to know the status of p_3 at all. Consider the derivative of C with respect to p_1 and p_2, $\partial C/\partial p_1 = C_{p_1} \oplus C_{\bar{p}_1} = 1$; $\partial C/\partial p_2 = C_{p_2} \oplus C_{\bar{p}_2} = 1$, so all changes of p_1 and p_2 are needed for evaluating C. Therefore, the back-end needs to be aware of every change of p_1 and p_2.   ■

Additionally, it is interesting to note that if a predicate is redundant for all the rules in the given back-end rule repository, we can disregard that predicate and hence, the corresponding observations completely.

2.4.4 Optimizations Based on Predicate Value Analysis

The number and size of transmissions can be reduced further by analyzing the values of the observed predicates which are co-observed by a particular information observer. Instead of sending the individual values of all predicates, we can merge the values of some predicates either by ANDing or by ORing, depending on the rule semantics. The predicates amenable to this optimization should have the following properties:

- The derivative of a rule with respect to the predicate must not be 1. This is because, if the derivative of a rule with respect to the predicate is 1, then the value of the predicate is critical for the back-end. Hence, the back-end needs to know every change of the predicate separately.
- Either all or none of them are part of a rule.
- One co-factor for each predicate is constant.
  - If either of the co-factors of the predicates is 1, then instead of sending their individual values, the observer can send only one value by ORing them.
  - If either of the co-factors of the predicates is 0, then instead of sending their individual values, the observer can send only one value by ANDing them.

Optimization Observation 4 If a set of co-observed predicates (observed at the same information observer) appear in the back-end rules in the same combining pattern or do not appear at all, then instead of sending their individual values, the information observer can transmit their combined value.   ∎

Example 9 Let $C_1 = p_1 + p_2 + p_3$, $C_2 = p_3 + p_4$ be two rules, and consider that p_1 and p_2 are co-observed and transmitted by the same information observer. In this case, p_1 and p_2 can be combined using the OR operator before transmission. However, p_3, p_4 cannot be combined, since p_3 is associated with the rule C_1 but not p_4. The value of p_3 is required to evaluate C_1. To combine a set of co-observed predicates before transmission, either all of them need to be present in a rule in the same combining pattern (AND/OR) or all of them have to be absent from the rule.

Consider the second rule in Sect. 2.2, i.e., $C_2 = p_5 + p_6 + p_7.p_8$ and let us assume p_5, p_6, p_7, and p_8 are co-observed. If an information observer needs to send the status of p_5 and p_6, then instead of sending their individual values, it is better to send the value of $p_5 + p_6$, because the rule is true if any one of them is true, if both of them are false, then the decision of the rule depends on the value of both p_7 and p_8. Similarly, the values of p_7 and p_8 can be combined using AND. ∎

2.4.5 Optimizations Based on Interest Set Transmission

The optimizations discussed above can be considered to be done at initialization time by processing the rules and appropriately passing on the relevant information to the observers. Additionally, at run-time, the back-end may send further updates as and when necessary, with a message to the information observers to abstain from observing/transmitting information, depending on the current snapshot, with a note that they will be polled for further updates as needed. The information observers take this information over and above the information they had at the initialization step, and this leads to further reductions in transmissions as we have explained above.

We now delve on some specific run-time strategies that can lead to further reductions by selectively filtering on the predicates to be observed or transmitted, depending on the rule semantics. All changes of the predicates belonging to the set of critical predicates are required to be transmitted always. However, for the set of non-critical predicates, this is not the case and we have a chance of a further optimization, as we describe below. Depending on the current snapshot, a rule may need one of the multiple sets of data predicates on which it depends, which we formalize as the notion of the *interest set* below.

Definition 10 (Interest Set) The interest set $\mathcal{I}_c$ of a rule C is a set of predicates which is sufficient to evaluate the truth value of the rule based on the current snapshot. ∎

Example 10 Consider the rule $C = p_1 + p_2.p_3 + p_4$ and a snapshot ($p_1 = 1$, $p_2 = 1$, $p_3 = 1$, $p_4 = 0$). To evaluate C, any of the sets $\{p_1\}$, $\{p_2, p_3\}$ is sufficient to take the decision. Hence, the *interest set* of C consists of $\{p_1\}$ and $\{p_2, p_3\}$. ∎

An arbitrary choice of the interest set may end up in a situation where a large number of transmissions of the data predicates take place though the size of the optimal set is quite smaller than the size of the chosen set. For a rule, there may exist more than one interest set. We now define the concept of a minimal interest set.

Definition 11 (Minimal Interest Set) Minimal interest set of a set of rules is the minimal set of predicates that is sufficient to evaluate all the rules. ∎

The minimal interest set may not be always unique. The minimal interest set can again be computed at initialization stage for the given set of rules and distributed to the information observers for judicious discrimination of what to transmit at what points. This proposal may, however, need a substantial resource requirement at the information observer end, and may not always be achievable in practice. In contrast, we propose here, how to compute the minimal interest sets based on a value snapshot, such that the back-end can communicate the necessary information to the information observers and poll for information as and when needed.

Example 11 Consider the following set of rules $C_1 = p_1 + p_2 + p_3, C_2 = p_2 + p_4 + p_5, C_3 = p_4 + p_6$ and a snapshot $p_1 = 1, p_2 = 1, p_3 = 0, p_4 = 1, p_5 = 0, p_6 = 1$. Some of the interest sets that can help us evaluate the rules are $\{p_1, p_2, p_4\}, \{p_1, p_2, p_6\}, \{p_2, p_4\}, \{p_1, p_4\}, \{p_2, p_6\}$. The minimal interest set in this case is one of the following sets $\{p_2, p_4\}, \{p_1, p_4\}, \{p_2, p_6\}$. From the example, we can see that the minimal interest set is not unique. ∎

Algorithms 1 and 2 present the strategy for minimal interest set selection. The idea is to compute the interest sets of each rule and then select the minimum combination. Analyzing the structure of the rules, we form a Boolean formula that captures the requirements of each rule and use a Boolean satisfiability solver to extract all satisfying solutions. Any of the solutions corresponds to an interest set and we choose the one with the minimal cardinality.

Algorithm 1 RunTimeInterestSetSelection

1: Input: Current Predicate Snapshot, Set of rules $\hat{C}$
2: Output: $\mathcal{U}$ ▷ *Set of predicates needed by the back-end*
3: $Q = \hat{C}$.
4: **for** $c \in Q$ **do**
5:    Compute the interest set of c.
6: **end for**
7: All predicates in P_1 (the set of critical predicates) are included in $\mathcal{U}$.
8: If the interest set of a rule contains only one component, all elements of that set are included in $\mathcal{U}$ and the corresponding rule is removed from Q.
9: If an interest set of a rule $c \in Q$, contains a component whose elements are already included in $\mathcal{U}$, then the corresponding rule is removed from Q.
10: MinimalInterestSetSelection($Q, \mathcal{U}$)

Algorithm 2 MinimalInterestSetSelection

1: Input: $Q, \mathcal{U}$
2: Output: $\mathcal{U}$ ▷ *Set of predicates needed by the back-end*
3: **for** $c \in Q$ **do**
4:    Construct a Boolean formula ($\mathcal{F}$) as follows:
5:    For each predicate p, a variable x_p is defined as:

$$x_p = 0, \text{ if } p \text{ is being considered}$$
$$\text{as an element of } \mathcal{U}$$
$$= 1, \text{ otherwise}$$

6:    Combine each Boolean variable corresponding to each member of every interest set of c using AND to get a product term.
7:    If one of the data predicates in a set is already included in $\mathcal{U}$, then the value of its corresponding Boolean variable is always 1, hence we can remove the Boolean variable from the formula.
8:    All product terms in one interest set are combined using OR operator to get a sum of product term.
9:    All sum of product terms corresponding to c are combined using AND operator. i.e., $\mathcal{F} = \mathcal{F}$ AND (sum of product term corresponding to c).
10: **end for**
11: Get all possible satisfiable solutions of the formula $\mathcal{F}$. Choose the solution whose cardinality of number of 1s is minimum. In case of tie, break it arbitrarily. The corresponding data predicates in $\mathcal{U}$ are included.

Example 12 pt In this example, we show how the algorithm MinimalInterestSet Selection works. Let, $C_1 = p_1 + p_2.p_3 + p_4$; $C_2 = p_2 + p_5$; $C_3 = p_3 + p_6.p_7$ be three rules. Let the current snapshot be ($p_1 = 1, p_2 = 1, p_3 = 1, p_4 = 0, p_5 = 1, p_6 = 1, p_7 = 1$). So, the interest set of C_1 is $\{p_1\}, \{p_2, p_3\}$, the interest set of C_2 is $\{p_2\}, \{p_5\}$, and the interest set of C_3 is $\{p_3\}, \{p_6, p_7\}$. After step 7 of Algorithm 2, the Boolean formula obtained is: $\mathcal{F} : (x_{p_1} \vee (x_{p_2} \wedge x_{p_3})) \wedge (x_{p_2} \vee x_{p_5}) \wedge (x_{p_3} \vee (x_{p_6} \wedge x_{p_7}))$. Consider all satisfiable solutions of $\mathcal{F}$ and we choose the solution $(x_{p_1}, x_{p_2}, x_{p_3}, x_{p_4}, x_{p_5}, x_{p_6}, x_{p_7}) :: (0, 1, 1, 0, 0, 0, 0)$ and include p_2, p_3 in $\mathcal{U}$. This is the minimal set of predicates the back-end needs to know in this case.   ■

The optimization strategies do not necessarily give us the optimal set required to evaluate all the rules defined at the back-end, as we already discussed in the comparison between unateness analysis and multiple variable co-factor analysis. However, our algorithm for minimal interest set selection gives us the minimal set among all the sets that we can construct after applying the optimizations 1,2,3, and 4. This is expressed by the following lemma.

Lemma 1 *Algorithm 2 gives a minimal interest set.*   ∎

Proof. We prove this by contradiction. Let us assume that there exists a different minimal interest set $\mathcal{U}' \neq \mathcal{U}$. The case where the cardinality of $\mathcal{U}$ is equal to the cardinality of $\mathcal{U}'$ is trivial. We consider the case where $|\mathcal{U}'| < |\mathcal{U}|$. If $\mathcal{U}'$ is the minimal interest set, then it gives a solution to the Boolean formula $\mathcal{F}$ generated in the MinimalInterestSetSelection procedure. Therefore, the number of 1s for this solution is less than that of our solution, which is contrary to the fact that we choose the solution where the cardinality of number of 1s is minimum. Hence, $\mathcal{U}$ is the minimal interest set.   ∎

2.4.6 Optimizations Based on Rule Clustering

A further optimization can be done for handling multiple rules. The basic idea is to have multiple Boolean formulae for each *Rule Family* formed by clustering the rules, instead of dealing with each in isolation. Then all these formulae can be evaluated together. Rule families are mutually exclusive and collectively exhaustive. The data predicate set corresponding to the rule families are pairwise disjoint. We now propose an algorithm for generating the rule families in Algorithm 3.

Algorithm 3 GenerateRuleFamily

1: Input: $\mathcal{C} = \{\mathcal{C}_1, \mathcal{C}_2, \ldots, \mathcal{C}_k\}$
2: Output: $\{\mathcal{C}_{F_1}, \mathcal{C}_{F_2}, \ldots, \mathcal{C}_{F_m}\}$
3: Initialize $i = 1$; $p = p_1$. A flag variable is set to false for each rule and each data predicate.
4: Initialize $\mathcal{C}_{F_1} = \phi$; $\mathcal{P}_{F_1} = p_1$
5: **repeat**
6:     $\mathcal{C}_{F_i} = \mathcal{C}_{F_i} \cup \{\ \mathcal{C}_j$ such that $\partial \mathcal{C}_j / \partial p \neq 0\}$
7:     Mark the flag of p as true.
8:     Revise the set $\mathcal{P}_{F_i}$. Include all data predicates associated with each rule $c \in \mathcal{C}_{F_i}$ and not already added in $\mathcal{P}_{F_i}$.
9:     Mark flag of each rule $c \in \mathcal{C}_{F_i}$ as true.
10:    **if** $\mathcal{P}_{F_i}$ contains any predicate p_j with flag as false **then**
11:        $p = p_j$
12:    **else**
13:        $i = i + 1$;
14:        $p = $ A data predicate whose flag is false; If no such p exists, terminate the procedure.
15:    **end if**
16: **until** (All predicates or all rules are examined)

Example 13   Consider a set of five rules $C_1 = p_1 + p_2 \cdot p_3 + p_4, C_2 = p_1 + p_5, C_3 = p_5 + p_6, C_4 = p_7 + p_8$, and $C_5 = p_7 + p_9$. The rule families are $\mathcal{CF}_1 = C_1, C_2, C_3$, $\mathcal{CF}_2 = C_4, C_5$. For each family, we construct the Boolean formula individually to get the number of predicates for evaluating all the rules belonging to that rule family. ∎

The intuitive idea of Algorithm 3 is to generate rule families which are mutually exclusive and collectively exhaustive. Since the set of data predicates and the set of rules are finite, the algorithm terminates. We can generate the rule families as part of the derivative analysis step as well. However, the main bottleneck for this procedure is the satisfying solution generation step, which is computationally expensive and hence, depending on the power of the computation back-end, may not be always efficient to perform at run-time. To make things faster, our final proposal is based on prediction, which may not give us an optimal solution but within a desired time limit, it is able to give us a good solution as we establish through our experiments.

2.4.7   Optimizations Using Prediction

The first few steps in this approach are same as above. The interest set of all the rules in $\hat{C}$ are computed at first. The final set $\mathcal{U}$ of the predicates that need to be captured by the information flow processing (IFP) engine, generated in each round, contains all the critical predicates and the predicates from the interest set corresponding to the rules having singleton interest sets satisfying the current snapshot at that round. Rules for which at least one interest set is considered in the final set $\mathcal{U}$ are ruled out, the remaining rules participate in this prediction algorithm. In other words, the rules that participate in this prediction algorithm have two or more interest sets, and the decision to select one of them is where our optimization proposal comes in. We use a simple predicate history outcome-based prediction strategy. We wish to select interest sets which remain unchanged for a long time, and thereby, can lead to more transmission savings. Therefore, for each rule we choose an interest set consisting of predicates that change less frequently. This leads to a further optimization in our methodology. However, it can be noted that we do not lose out on any information or rule evaluation. If the predicted set does change at run-time, we poll the remaining information observers for the other predicates.

Example 14   Consider a rule and its corresponding action as follows: Send an alert to the fire alarm if the room temperature is greater than 40 °C and smoke is detected in the room. In this example, the second event (smoke detection) is likely to change less frequently than the first one. ∎

In our prediction-based strategy, when a given rule has two interest sets containing two different predicates, we choose the interest set containing the event that has less chances of changing based on history. We now discuss the prediction strategy. For each predicate, we store a history h at the back-end, where h denotes the number of changes (0 to 1 or 1 to 0 is considered as change) that the back-end has noticed over the history of computation. A lower value of h implies the predicate changes less frequently. An interest set usually contains one or more predicates. In an interest set, predicate(s) with low history values are considered as the representative predicates for that interest set. While selecting an interest set for each rule, we select the one whose representative predicate has lowest history value. As is intuitively obvious, the final set $\mathcal{U}$ that is generated in each round using prediction may not be optimal in terms of cardinality. At the initial instant, the information observers send all the predicates whose derivatives are not 0 to the back-end engine. Next time onwards, the information observers send the predicate to the IFP engine only if there is a change in the value of the predicate and the back-end polls for its value, based on the minimal interest set computation step. The correctness of the above scheme follows from the fact that we choose the predicates from the interest set.

2.5 Case Study

Figure 2.2 summarizes the flow of our framework. In this section, we present two detailed case studies illustrating some of the step-by-step optimizations.

Fig. 2.2 Workflow

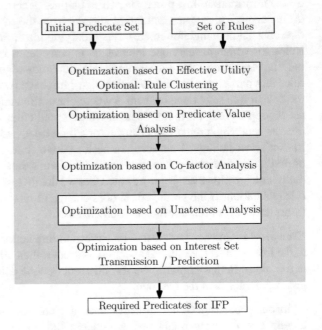

2.5.1   A Large-Scale Weather Monitoring System

Consider a weather monitoring system where the information observers transmit the following predicates:

- p_1: Weather cloudy;
- p_2: Precipitation $>80\%$;
- p_3: Temperature $\geq 25\,°C$;
- p_4: It is raining;
- p_5: Water-level $\geq$ danger level (L_1);
- p_6: Drainage system is not working properly;
- p_7: Wind speed $>50\,m/s$;
- p_8: Snow melting rate $\geq$ threshold (t_1);
- p_9: Humidity $\geq 90\%$.

The information observers and their predicates are as follows:

- s_1 observes the weather (p_1) and the wind speed (p_7).
- s_2 observes the temperature (p_3).
- s_3 observes precipitation (p_2).
- s_4 observes whether it is raining (p_4).
- s_5 observes water-level (p_5) and the drainage system (p_6).
- s_6 observes snow melting rate (p_8) and humidity (p_9).

The rules and the rule-triggered actions are as follows:

- Send rain alert if any of the following condition is true:

  - The weather is cloudy and Precipitation $>80\%$
  - The weather is cloudy and Temperature $\geq 25\,°C$.

  In our representation, we express this as:
  $p_1 \cdot p_2 + p_1 \cdot p_3$: Send rain alert ($\mathcal{A}_1$);
- Send snow melting alert if the temperature $\geq 25\,°C$
  p_3: Send snow melting alert ($\mathcal{A}_2$);
- Send unpleasant weather alert if any of the following condition is true:

  - The temperature $\geq 25\,°C$
  - Humidity $\geq 90\%$.

  We express this as: $p_3 + p_9$: Send unpleasant weather alert ($\mathcal{A}_3$);
- Send flood alert if any of the following condition is true

  - It is raining and the water-level $\geq L_1$.
  - It is raining and the drainage system is not working properly.
  - It is raining and the water-level $\geq L_1$, the drainage system is not working properly and wind speed $>50\,m/s$.
  - Snow melting rate $\geq$ threshold (t_1).

  We express this as: $p_4 \cdot p_5 + p_4 \cdot p_6 + p_4 \cdot p_5 \cdot p_6 \cdot p_7 + p_8$: Send flood alert ($\mathcal{A}_4$);

In summary, we have the following rules and actions:

- $C_1 : p_1.p_2 + p_1.p_3 : \mathcal{A}_1$
- $C_2 : p_3 : \mathcal{A}_2$
- $C_3 : p_3 + p_9 : \mathcal{A}_3$
- $C_4 : p_4.p_5 + p_4.p_6 + p_4.p_5.p_6.p_7 + p_8 : \mathcal{A}_4.$

Transmission Optimization We explain our optimizations on the above example.

1. Derivative analysis:
   The result of the derivative analysis of the rules is shown in Table 2.5. As discussed earlier, we generate the rule families in the derivative analysis step itself. For this example, we have the following rule families:

   - $CF_1 = \{C_1, C_2, C_3\}$; corresponding predicate set $PF_1 = \{p_1, p_2, p_3, p_9\}$
   - $CF_2 = \{C_4\}$; corresponding predicate set $PF_2 = \{p_4, p_5, p_6, p_8\}$.

   Decision:

   - $P_1 = \{p_3\}$; crucial predicates.
   - $P_2 = \{p_7\}$; predicates not required for any rule.
   - $P_3 = \{p_1, p_2, p_4, p_5, p_6, p_8, p_9\}$; required predicates, not including the crucial predicates.

2. Predicate Value Analysis:
   s_6 handles the observation/transmission of the state of the water-level (p_5) and the drainage system (p_6). Only the rule C_4 uses the predicates p_5 and p_6. Instead of sending p_5 and p_6, s_6 can send the combined value as $p_5 + p_6$, which is represented as p_{5-6} below.
   The new set of required predicates:
   $P_4 = \{p_1, p_2, p_4, p_{5-6}, p_8, p_9\}$.
3. Co-factor analysis:
   Single variable and double variable co-factor analyses of the rules are shown in Tables 2.6 and 2.7, respectively.

Table 2.5 Result of derivative analysis

Derivative	Result
$\partial C_1/\partial p_1$	Not constant
$\partial C_1/\partial p_2$	Not constant
$\partial C_1/\partial p_3$	Not constant
$\partial C_2/\partial p_3$	1
$\partial C_3/\partial p_3$	Not constant
$\partial C_3/\partial p_9$	Not constant
$\partial C_4/\partial p_4$	Not constant
$\partial C_4/\partial p_5$	Not constant
$\partial C_4/\partial p_6$	Not constant
$\partial C_4/\partial p_7$	0
$\partial C_4/\partial p_8$	Not constant

Table 2.6 Single variable co-factor analysis

Positiveco-factor	Resultingset	Negativeco-factor	Resultingset
C_{1p_1}	$\{p_2, p_3\}$	$C_{1\bar{p}_1}$	0
C_{1p_2}	$\{p_1\}$	$C_{1\bar{p}_2}$	$\{p_1, p_3\}$
C_{1p_3}	$\{p_1\}$	$C_{1\bar{p}_3}$	$\{p_1, p_2\}$
C_{3p_3}	1	$C_{3\bar{p}_3}$	$\{p_9\}$
C_{3p_9}	1	$C_{3\bar{p}_9}$	$\{p_3\}$
C_{4p_4}	$\{p_{5-6}, p_8\}$	$C_{4\bar{p}_4}$	$\{p_8\}$
$C_{4p_{5-6}}$	$\{p_4, p_8\}$	$C_{4\bar{p}_{5-6}}$	$\{p_8\}$
C_{4p_8}	1	$C_{4\bar{p}_8}$	$\{p_4, p_{5-6}\}$

Table 2.7 Double variable co-factor analysis

Co-factor	Resulting set
C_{1p_1,p_2}	1
C_{1p_1,p_3}	1
$C_{1\bar{p}_2,\bar{p}_3}$	0
$C_{3\bar{p}_3,\bar{p}_9}$	0
$C_{4p_4,p_{5-6}}$	1
$C_{4\bar{p}_4,\bar{p}_8}$	0
$C_{4\bar{p}_{5-6},\bar{p}_8}$	0

Table 2.8 Unateness analysis

Rule	Positive unate with	Negative unate with
C_1	$\{p_1, p_2, p_3\}$	–
C_3	$\{p_3, p_9\}$	–
C_4	$\{p_4, p_{5-6}, p_8\}$	–

4. Unateness analysis:
   The rules and the predicate list on which the rule is positive or negative unate is shown in Table 2.8. Using the pre-processing information stored in Tables 2.5, 2.6, 2.7, and 2.8, the back-end takes decision at run-time, which set of predicates are required to evaluate all the rules.

5. Optimization based on interest set:
   We consider some random changes of the predicates as in Table 2.9. Back-end computation at the initial instance: The observers transmit the initial values of the predicates (1, 1, 1, 0, 1(value for p_{5-6}), 0, 1) to the back-end. Since p_3 is a crucial predicate, the back-end needs all the changes of p_3. p_4, p_8 are enough to evaluate C_4. Using p_3, the back-end can evaluate C_2, C_3. In order to evaluate C_1, the back-end generates the Boolean formula: $\mathcal{F} = p_1.p_2 + p_1$ (since p_3 is already considered); solving the formula, we can infer the following predicates as the currently required set: $\mathcal{U} = \{p_1, p_3, p_4, p_8\}$;

Table 2.9 Random changes of the predicates

Cycle	p_1	p_2	p_3	p_4	p_5	p_6	p_7	p_8	p_9
1	1	1	1	0	1	1	0	0	1
2	1	0	1	0	0	0	1	0	0
3	1	0	1	0	1	0	0	0	1
4	0	0	1	0	1	1	1	0	0
5	0	1	1	0	0	1	0	0	1

Back-end computation in the next instances:

In the fourth instance the value of p_1 changes. Now p_1 is enough to evaluate C_1. So the predicates needed for the back-end: $\mathcal{U} = \{p_1, p_3, p_4, p_8\}$. Comparing the transmission profiles, we have

- No. of transmissions in usual synchronous case: 45
- No. of transmissions in usual asynchronous case: 25
- No. of transmissions in our case: 8.

2.5.2   Health Monitoring System: A Medical CPS Context

Consider a body sensor network [17], where a set of sensors $S_1 \ldots S_5$ monitor the parameters mentioned below for a person and transmit data to a centralized controller.

S_1: Heart rate (HR), ECG; S_2: Oxygen saturation (SpO_2), Respiratory rate (RR); S_3: Body temperature (BT); S_4: Blood pressure (Bp: (systole (Sys), diastole (Dia))), Body glucose level (BGL); S_5: Activities (e.g., sitting, standing, lying, walking), body mass index (BMI).

Consider the following predicates:

$p_1 : BT > 99°F$, $p_2 : Sys > 140$, $p_3 : Sys < 110$, $p_4 : SpO_2 < 90$, $p_5 : HR > 100$, $p_6 : RR > 20$ (for adults), $p_7:BMI \geq 30$, p_8: Abnormal ECG, p_9: Person is resting (i.e., lying down or sitting or standing) $p_{10} : BGL > 200$

We have the following rules:

- $C_1 : p_9.p_5.p_6$: Alert for atrial fibrillation/asthma (A_1);
- $C_2 : p_1 + (p_2.p_9) + p_3 + (p_6.p_9)$: Alert for weakness ($A_2$);
- $C_3 : (p_2.p_4.\bar{p}_7) + (p_9.p_5) + p_8$: Alert for cardiac problem (A_3);
- $C_4 : p_4 + (p_9.p_6)$: Alert for lung problem (A_4);
- $C_5 : p_{10}$: Alert for diabetes (A_5).

Table 2.10 presents the positive and negative co-factors of the rules. Consider a random snapshot of the predicates for an interval of five time instants, as shown in Table 2.11. We now illustrate our different steps below. Clearly we have two different clusters $\{C_1, C_2, C_3, C_4\}, \{C_5\}$.

The sensors transmit the initial values of the predicates (0, 0, 1, 0, 0, 0, 0, 0, 0, 0) to the back-end. The back-end needs all the changes of p_{10}, being the only predicate in C_5. p_3 is the only predicate required to evaluate C_2. In order to

Table 2.10 Co-factors of the rules

Rule	Positive co-factor	Negative co-factor
C_1	$\{p_5, p_6, p_9\}$	$\{\bar{p}_5\}, \{\bar{p}_6\}, \{\bar{p}_9\}$
C_2	$\{p_1\}, \{p_2, p_9\}, \{p_3\}, \{p_6, p_9\}$	$\{\bar{p}_1, \bar{p}_2, \bar{p}_3, \bar{p}_6\}, \{\bar{p}_1, \bar{p}_3, \bar{p}_9\}$
C_3	$\{p_2, p_4, \bar{p}_7\}, \{p_9, p_5\}, \{p_8\}$	$\{\bar{p}_2, \bar{p}_9, \bar{p}_8\}, \{\bar{p}_4, \bar{p}_9, \bar{p}_8\}, \{p_7, \bar{p}_9, \bar{p}_8\}$
		$\{\bar{p}_2, \bar{p}_5, \bar{p}_8\}, \{\bar{p}_4, \bar{p}_5, \bar{p}_8\}, \{p_7, \bar{p}_5, \bar{p}_8\}$
C_4	$\{p_4\}, \{p_9, p_6\}$	$\{\bar{p}_4, \bar{p}_6\}, \{\bar{p}_4, \bar{p}_9\}$
C_5	$\{p_{10}\}$	$\{\bar{p}_{10}\}$

Table 2.11 Changes on the predicates

Time instant	p_1	p_2	p_3	p_4	p_5	p_6	p_7	p_8	p_9	p_{10}
1	0	0	1	0	0	0	0	0	0	0
2	0	0	1	0	1	1	0	1	0	0
3	0	0	1	0	0	0	0	0	0	0
4	1	0	1	0	1	1	0	1	0	0
5	1	1	1	0	0	0	0	0	1	0
Total	2	2	1	1	5	5	1	5	2	1

evaluate the remaining rules, the back-end generates the Boolean formula: $\mathcal{F} = (x_{p_5} \lor x_{p_6} \lor x_{p_9}) \land (x_{p_2} \lor x_{p_4}) \land (x_{p_5} \lor x_{p_9}) \land x_{p_8} \land x_{p_4} \land (x_{p_6} \lor x_{p_9})$; solving the formula, we can infer the following predicates as the currently required set: $U = \{p_3, p_4, p_8, p_9, p_{10}\}$. In the second instance, the value of p_8 changes and therefore only the value of p_8 is transmitted in this cycle. This causes C_3 to evaluate to true. The new Boolean formula is therefore, $\mathcal{F} = (x_{p_5} \lor x_{p_6} \lor x_{p_9}) \land x_{p_4} \land (x_{p_6} \lor x_{p_9})$; solving the formula, we can infer the minimal interest set in the current cycle: $U = \{p_3, p_4, p_8, p_9, p_{10}\}$, which is same as above. In the third instance, the value of p_8 changes and therefore only the value of p_8 is transmitted in this cycle. The new Boolean formula is therefore same as in the first cycle. The minimal interest set in this cycle: $U = \{p_3, p_4, p_8, p_9, p_{10}\}$ is also same as earlier. In the fourth instance also, the value of p_8 changes and therefore the value of p_8 is transmitted in this cycle. The new Boolean formula remains same as the second instance. The set of required predicates in this cycle, $U = \{p_3, p_4, p_8, p_9, p_{10}\}$, is therefore exactly same as earlier. In the fifth instance, the values of p_8, p_9 change and therefore only the values of p_8, p_9 are transmitted in this cycle. As p_9 changes, the values of p_5 and p_6 are also required in this cycle, since our observation was that as long as p_9 remains same, the changes of p_5 and p_6 are not required. The new Boolean formula is therefore, $\mathcal{F} = (x_{p_5} \lor x_{p_6}) \land (x_{p_2} \lor x_{p_4}) \land x_{p_5} \land x_{p_8} \land x_{p_4} \land x_{p_6}$; solving the formula, we can infer the minimal interest set in fifth cycle: $U = \{p_3, p_4, p_5, p_6, p_8, p_{10}\}$.

While we have 25 transmissions in the usual asynchronous case, we have 17 in our case, as shown in Table 2.12.

Table 2.12 Transmission profile

Time instant	p_1	p_2	p_3	p_4	p_5	p_6	p_7	p_8	p_9	p_{10}
1	0	0	1	0	0	0	0	0	0	0
2	X	X	X	X	X	X	X	1	X	X
3	X	X	X	X	X	X	X	0	X	X
4	X	X	X	X	X	X	X	1	X	X
5	X	X	X	X	0	0	X	0	1	X
Total	1	1	1	1	2	2	1	5	2	1

2.6   Experimental Result

We implemented the proposed framework in Java. We tested our method on random environments, where the number of predicates, the rules, and the number of information observers and their observation sets were randomly generated. The results are shown in Table 2.13. We consider a total window of 15 snapshots over time. We present the effect of different optimization procedures through our experiments. Each row of the table corresponds to the results obtained on one experimental dataset. Column 1 and Column 2 of Table 2.13 represent the number of randomly generated predicates and rules, respectively. Column 3 shows the number of different clusters found in the rule set. Column 4 shows the number of predicates required after derivative analysis, whereas Column 5 indicates the number of critical predicates present in the rule set. We compare our method with the asynchronous transmission scheme with respect to the number of transmissions. We also present comparative results on our proposal with and without run-time optimizations. Column 6 presents the number of transmissions in a normal asynchronous transmission scheme (usual case), Column 7 shows the number of transmissions without the interest set optimization (Case 1), Column 8 presents the number of transmissions with minimal interest set selection (Case 2), and finally Column 9 shows the number of transmissions using prediction (Case 3). As can be seen from the results, the number of transmissions reduces with our optimizations (Cases 1 and 2). In some cases, the reduction is substantial, whereas in some cases, it is comparable. The bottom line is that these optimizations can lead to the same performance as in the asynchronous method in the worst case if the rule set does not lend itself to any of the optimizations we propose in this work. It is guaranteed theoretically and also as seen through the experiments, that our method (Case 1 and Case 2 without prediction) does not transmit more than the asynchronous method. However, with the prediction optimization, that is not always the case, as shown in Case 3. The prediction step being non-deterministic may or may not yield the desired improvement. A graphical view of the comparative number of transmissions is shown in Fig. 2.3, with increase in the number of predicates.

Table 2.13 Test-case statistics (#Tx = Number of transmissions, #P = Number of predicates, $\#P_\delta$ = Number of predicates after derivative analysis)

#P	No. of rules	No. of clusters	$\#P_\delta$	No. of critical predicates	#Tx (usual case)	#Tx (Case 1)	#Tx (Case 2)	#Tx (Case 3)
13	43	3	13	2	645	645	91	645
14	6	1	14	0	716	404	398	663
15	2	2	15	0	737	144	144	326
28	28	1	28	0	1420	1111	957	1276
28	7	2	28	2	1425	413	414	974
31	6	3	28	1	1549	239	239	902
35	8	1	35	0	1736	440	426	1207
36	39	2	36	2	1761	1425	1117	1761
38	31	2	38	3	1848	1198	1099	1692
41	9	1	40	0	2001	442	428	1071
44	31	1	43	1	2127	1466	1175	1988
44	66	2	44	8	2222	1890	1665	2133
46	38	2	46	3	2247	1843	1554	2102
49	103	2	49	9	2416	2367	2208	2416
51	26	2	50	5	2540	1348	1199	2134
60	68	1	60	3	3043	2539	2227	3043
60	27	5	59	6	3013	1306	1149	2507
63	24	3	62	5	3107	1707	1515	2417
64	126	2	64	7	3120	3071	2932	3120
68	68	2	68	7	3340	2424	2079	3242
73	40	2	73	2	3562	2120	1895	2959
73	33	3	72	4	3572	1971	1830	2947
74	77	2	74	12	3651	2470	2113	3494
76	107	2	76	9	3762	3505	3078	3669
78	75	2	78	6	3979	3208	2589	3871
79	34	3	78	4	3888	1702	1655	3095
83	58	2	83	4	4104	2874	2587	3780
87	92	2	87	8	4335	3290	2876	4131
98	116	2	98	13	4802	3728	3326	4563

2.7 Related Work

To the best of our knowledge, our proposed model is the first work of its kind and we have not found any work on data flow processing using rule analysis. A lot of work has been done in data flow processing domain. Data stream processing model [6] and complex event processing model [7] are two main directions in this research area. The applications of IFP come from different domains, e.g., environmental monitoring system [18], financial applications like continuous stock analysis [19], network traffic [20], health monitoring [1], fraud detection [21], etc.

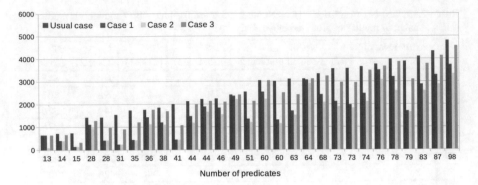

Fig. 2.3 Number of predicates vs number of transmissions

It is worth mentioning that public alert services are becoming popular in smart cities. Context-awareness enhances the efficiency of such a system. Banerjee et al. [22] proposes a framework where a selective notification will improve the pervasive experience. Stream reasoning is the backbone of this system which uses rule-based reasoning and queries. The paper presents a context-aware access control system fitted to ubiquitous medical sensor networks. Nath [23] proposes a middleware (acquisitional context engine, ACE) for auto-learning the relationships among various context attributes and using them to optimize inference caching and speculative sensing. A multidimensional framework for context-aware systems is proposed in [24]. Olaru et al. [25] propose a software agent-based model for a middleware on ambient intelligence (AmI). Here, context-awareness is implemented both in "agent's representation of context information" and in the "logical topology of the agent system." The limitation of this model is the orientation towards decentralization of the system and reliability on local behavior. A scalable and energy-efficient context monitoring twofold framework has been proposed in [26] for sensor-rich mobile environments where limited resource is concerned. The idea consists of continuous detection of context changes followed by bi-directional approaches to the context monitoring problem. The author claims to obtain higher efficiency in energy consumption. Other energy-efficient data transmission reduction methodologies [27, 28] are proposed for periodical data gathering in WSN. These methods exploit overhearing, where the redundancy of the reading of each node in a WSN is determined by the overheard packets transmitted by its neighbors and restricts the redundant reading transmission. Ferrigno et al. [29] proposed a method for reducing the energy consumption on a visual sensor node in wireless sensor networks (WSN).

Many proposals in literature explore energy-efficient transmission schemes. Ciullo et al. [30] proposed an energy-efficient scheme for data collection in WSN using mobile elements (with mobility control). Here, for minimizing "total transmission energy" within a certain "travel time," they considered the problem of "optimal vehicle trajectories" as a function of "data at each sensor." For storing or sending bulky data, Chang et al. [31] used Bloom filters to reduce the memory and network transmission necessities. Using this approach, more information from WSN can be collected and also network lifetime can be extended. For efficient query response in a context-based system, the *RETE* algorithm [32] is used, which basically deals with a set of *If-then (-else)* rules. In this approach, facts are matched against rules. We also use this kind of a rule to reduce the number of transmissions. The main difference of our work with existing literature is the fact that we leverage on the Boolean nature of the context rule for efficient message dissemination.

2.8 Conclusion

In this chapter, we present a novel architecture for an information flow processing engine. Experimental results show promising improvements in the number of transmissions. With information processing and sensing systems gaining widespread popularity in recent times, we believe our work will have interesting applications.

References

1. O. Garcia-Morchon, K. Wehrle, Modular context-aware access control for medical sensor networks, in *SACMAT* (2010), pp. 129–138
2. RK-Real-time-City.pdf [Online]. Available http://eprhtints.maynoothuniversity.ie/5625/1/RK-Real-time-City.pdf
3. F. Bonomi, R. Milito, J. Zhu, S. Addepalli, Fog computing and its role in the internet of things, in *Proceedings of the First Edition of the MCC Workshop on Mobile Cloud Computing* (ACM, New York, 2012), pp. 13–16
4. L. Filipponi et al., Smart city: an event driven architecture for monitoring public spaces with heterogeneous sensors, in *SENSORCOMM* (IEEE, Piscataway, 2010), pp. 281–286
5. G. Cugola, A. Margara, Processing flows of information: from data stream to complex event processing. ACM Comput. Surv. **44**(3), 15:1–15:62 (2012)
6. B. Babcock et al., Load shedding for aggregation queries over data streams, in *Proceedings of the 20th International Conference on Data Engineering*. ICDE (2004), p. 350
7. D.C. Luckham, *The Power of Events: An Introduction to Complex Event Processing in Distributed Enterprise Systems* (Addison-Wesley Longman Publishing, Boston, 2001)
8. Y. Xu et al., Prediction-based strategies for energy saving in object tracking sensor networks, in *2004 IEEE International Conference on Mobile Data Management, 2004. Proceedings* (IEEE, Piscataway, 2004), pp. 346–357
9. J. Yick et al., Wireless sensor network survey. Comput. Netw. **52**(12), 2292–2330 (2008)
10. D. Kossmann, The state of the art in distributed query processing. ACM Comput. Surv. (CSUR) **32**(4), 422–469 (2000)
11. S. Adali et al., Query caching and optimization in distributed mediator systems, in *ACM SIGMOD Record*, vol. 25, no. 2 (ACM, New York, 1996), pp. 137–146
12. S. Chattopadhyay, A. Banerjee, N. Banerjee, A data distribution model for large-scale context aware systems, in *Proceedings of the 10th International Conference on Mobile and Ubiquitous Systems: Computing, Networking and Services (MobiQuitous 2013)* (2013)

13. S. Chattopadhyay, A. Banerjee, B. Yu, An utility-driven data transmission optimization strategy in large scale cyber-physical systems, in *2017 Design, Automation & Test in Europe Conference & Exhibition* (2017)
14. G.D. Hachtel, F. Somenzi, *Logic Synthesis and Verification Algorithms*, 1st edn. (Kluwer Academic Publishers, Norwell, 2000)
15. M.M. Mano, *Digital Logic and Computer Design* (Dorling Kindersley, London, 2013)
16. S.B. Akers, Binary decision diagrams. IEEE Trans. Comput. **27**(6), 509–516 (1978)
17. O. Salem et al., Anomaly detection in medical wireless sensor networks using SVM and linear regression models. Int. J. E-Health Med. Commun. **5**(1), 20–45 (2014)
18. K. Broda et al., Sage: a logical agent-based environment monitoring and control system, in *Proceedings of the European Conference on Ambient Intelligence* (2009), pp. 112–117
19. A. Demers et al., Towards expressive publish/subscribe systems, in *Proceedings of the 10th International Conference on Advances in Database Technology*. EDBT'06 (2006), pp. 627–644
20. H. Debar, A. Wespi, Aggregation and correlation of intrusion-detection alerts, in *RAID* (2001), pp. 85–103
21. N.P. Schultz-Møller et al., Distributed complex event processing with query rewriting, in *ACM DEBS* (2009), pp. 4:1–4:12
22. S. Banerjee, D. Mukherjee, P. Misra, 'what affects me?': a smart public alert system based on stream reasoning, in *ICUIMC* (2013), pp. 22:1–22:10
23. S. Nath, Ace: exploiting correlation for energy-efficient and continuous context sensing, in *Proceedings of the 10th International Conference on Mobile Systems, Applications, and Services*. MobiSys '12 (2012), pp. 29–42
24. G. Fischer, Context-aware systems: the 'right' information, at the 'right' time, in the 'right' place, in the 'right' way, to the 'right' person, in *Proceedings of the International Working Conference on Advanced Visual Interfaces*. AVI '12 (2012), pp. 287–294
25. A. Olaru, A.M. Florea, A. Fallah Seghrouchni, A context-aware multi-agent system as a middleware for ambient intelligence. Mob. Netw. Appl. **18**(3), 429–443 (2013)
26. S. Kang, J. Lee, H. Jang, H. Lee, Y. Lee, S. Park, T. Park, J. Song, SeeMon: scalable and energy-efficient context monitoring framework for sensor-rich mobile environments, in *Proceedings of the 6th International Conference on Mobile Systems, Applications, and Services*. MobiSys '08 (2008), pp. 267–280
27. Y. Iima, A. Kanzaki, T. Hara, S. Nishio, Overhearing-based data transmission reduction for periodical data gathering in wireless sensor networks, in *International Conference on Complex, Intelligent and Software Intensive Systems, 2009. CISIS '09, March* (2009), pp. 1048–1053
28. P. Basu, J. Redi, Effect of overhearing transmissions on energy efficiency in dense sensor networks, in *IPSN '04: Proceedings of the third international symposium on Information Processing in Sensor Networks, April* (2004)
29. L. Ferrigno, S. Marano, V. Paciello, A. Pietrosanto, Balancing computational and transmission power consumption in wireless image sensor networks, in *Proceedings of the 2005 IEEE International Conference on Virtual Environments, Human-Computer Interfaces and Measurement Systems, 2005. VECIMS 2005, July* (2005), 6 pp.
30. D. Ciullo, G. Celik, E. Modiano, Minimizing transmission energy in sensor networks via trajectory control, in *2010 Proceedings of the 8th International Symposium on Modeling and Optimization in Mobile, Ad Hoc and Wireless Networks (WiOpt), May* (2010), pp. 132–141
31. M.L. Stephen Chang, A. Kirsch, Energy and storage reduction in data intensive wireless sensor network applications, in *Computer Science Group, Harvard University, Cambridge, MA*, vol. TR-15-07 (2007)
32. C.L. Forgy, Rete: a fast algorithm for the many pattern/many object pattern match problem, in *Expert Systems*, ed. by P.G. Raeth (IEEE Computer Society Press, Los Alamitos, 1990), pp. 324–341 [Online]. Available http://dl.acm.org/citation.cfm?id=115710.115736

Chapter 3
Advanced Communications in Cyber-Physical Systems

Taslim Arefin Khan, Suraiya Tairin, Mahmuda Naznin, Md Zakirul Alam Bhuyian, and A. B. M. Alim Al Islam

3.1 Introduction

The recent technological developments offer us a new generation of systems known as cyber-physical systems (CPSs). The emergence of CPSs introduces specialized networking and communication strategy, information technology, integrating them with physical world which enables the advancement of a new vision for the social facilities. A CPS is the integration of computation, communication, control, learning, and reasoning with physical processes. CPSs cannot be considered as conventional real-time systems or embedded systems. There are several features that exist in CPSs which make it different from other systems such as dynamically reconfigurable, fully automation, auto-assembly, and integration. A definition of CPSs was provided by Shankar Sastry from University of California, Berkeley in 2008 [1]:

> A cyber-physical system (CPS) integrates computing, communication and storage capabilities with monitoring and/or control of entities in the physical world, and must do so dependably, safety, securely, efficiently and real-time.

Cyber-physical systems provide a number of advantages. CPSs are safe and efficient engineered systems that control and integrate entities forming sophisticated systems with new competences and capabilities. CPSs can be applied extensively in several domains offering ample chances such as infrastructure control, energy control, environmental control, efficient transport system, tele-medicine, medical devices, assisted living, and agriculture. Complex systems having critical infrastruc-

T. A. Khan · S. Tairin · M. Naznin · A. B. M. A. Al Islam
Department of CSE, Bangladesh University of Engineering and Technology, Dhaka, Bangladesh

M. Z. A. Bhuyian (✉)
Department of Computer and Information Sciences, Fordham University, Bronx, NY, USA

© Springer Nature Switzerland AG 2020
S. Hu, B. Yu (eds.), *Big Data Analytics for Cyber-Physical Systems*,
https://doi.org/10.1007/978-3-030-43494-6_3

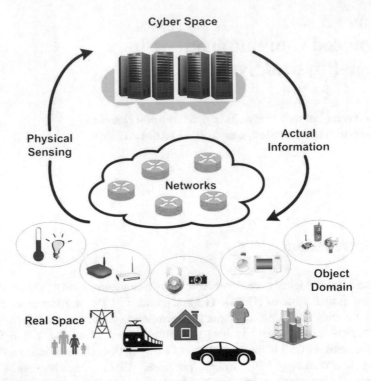

Fig. 3.1 Interconnection between cyber and physical objects [2]

ture such as water supply, gas production, electricity generation and distribution, and oil production are also outcome of CPSs.

The applications of CPSs expand from small systems (i.e., aircrafts) to very large systems (i.e., power grid). The interaction between cyber and physical object is shown in Fig. 3.1. Distinguished features of CPSs make it different from conventional wireless sensor network (WSN), desktop computing, and embedded systems. For example, the health care systems contain health information network, patient record, home care, operation management, hospitals and operating rooms, surgery and therapy, etc. Most of these tasks related to health care are highly controlled by computer systems with hardware and software components. Medical devices and systems need to be fully communicated and to act according to the patients' needs. Hence, medical devices are required to be dynamically reconfigured based on the circumstances of patients. For instance, devices such as infusion pumps, automatic oxygen delivery systems and sensors observing patient conditions need to be integrated into a new robust system to fulfil the patient needs. The main challenges in designing such CPS are maintaining safety, security, and reliability. Power management is another application of CPS which includes power electronics, power grid, and embedded control software integrated forming an efficient solution.

These types of CPS need specialization to ensure fault tolerance, security, safety, and decentralized control.

The research related to CPSs is still in embryonic stage. Currently, most of the research works are focused on specific domains such as networking, sensors, mathematics, computer science, system theory, and software engineering. For instance, complex systems are modeled utilizing diverse modeling methods, formalisms, and tools. A specific method or formalism is well suited to characterize either the cyber or the physical process but not both. CPSs are anticipated to be engineered systems that are versatile having robustness, self-organization, self-maintenance, autonomy, self-repair, efficiency, predictability, interoperability, global tracking and tracing, etc. There exist several challenges pertinent to CPSs that have to be addressed. The distinguished features of CPSs which make design of CPSs a challenging research issue are summarized below.

- Integration: CPSs integrate computation, control, and communication with physical world. Furthermore, some CPSs may have to integrate with other systems or other CPS [3]. For example, integration of CPS and cloud which may perform many tasks such as controlling power, storage, and data services [3].
- Limited capability: Some CPSs utilize devices with very limited capabilities and functionalities. Reason behind this is that the currently available devices may have limited capabilities. Besides, cost limitation is another reason. Such devices generally have limited computing, processing, communication, and storage capabilities.
- Heterogeneous devices: CPSs are conformed to several heterogeneous devices such as sensors, actuators, controllers, microcontrollers, networking devices, and communication devices. Moreover, these devices may operate at different location in different physical environments.
- Networks of different scales: CPSs include different types of networks such as wired/wireless network, Bluetooth, WLAN, mobile ad hoc network (MANET), and GSM in a distributive fashion. Besides, the scales of these networks and the types of devices are widely diversified.
- Power limitations: Some devices of CPS may be deployed in remote locations where no stable power sources are accessible. Therefore, the communication protocols of CPSs should be modeled considering the power limitations.
- Distributed control: Some CPS applications necessitate distributed control, processing, and decision making [3] to operate successfully. Furthermore, many applications require parallel processing for quick and prompt decision making.
- Real-time operations: CPSs often require to operate and take decision in real time. Besides, real-time operations also include real-time sensing, processing, communication, and response.
- Special communication: Some CPSs need specialized communication among different subsystems and the devices. Thus, communication among devices should be reliable and robust. More emphasis should be given in designing optimized communication techniques.

– Complexity regarding temporal and spatial scales: Most of CPSs may have to control multiple components at different time in different locations which make the networking paradigm of CPSs more complex.
– Dynamic adaptation: To cope up with different unpredictable environments, CPSs may have to reconfigure system settings. Hence, CPSs should have adaptive capabilities.
– Synchronization of control loops: CPSs are the outcome of collaboration man–machine where the information is circulated as loop from man to machine and machine to man. Therefore, CPSs should be integrated with advanced and synchronized feedback control technologies.
– Mobility: CPSs often include mobile devices which need proper synchronization to be connected with the rest of system. As mobile devices change locations frequently, specialized communication mechanism is needed to handle mobility in CPSs.
– Fault tolerance and reliability: CPSs applications are generally large-scale complicated embedded systems; hence, different types of fault should be detected and dealt with efficiently without hampering any regular operations of CPSs.
– Security and privacy: Most of CPSs involve distributed applications. Thus, the security and privacy of the information must be preserved.
– Context awareness: CPSs sometimes need to know the context of whole systems such as system status, locations of physical object to operate properly. Hence, proper synchronization and exchange of information are required for successful operation.
– Verification, validation, and certification: The relation between used methods and testing requires to be validated. The heterogeneous nature of CPS models needs compositional verification and testing methods.

Generally, a CPS integrates sensors, actuators, and controller with physical objects in large-scale. The operation of CPSs is usually divided among several subsystems. CPSs are considered as a form of wireless sensor and actuator networks (WSANs) [4, 5]. Here, sensors sense information about the physical world and actuators along with controllers process this information to take appropriate decisions. The performance of CPSs depends on the design and modeling of WSANs.

The nature of CPS totally relies on applications. Different applications may have different network architecture. For example, in a application of fire handling, sensors, actuators, and controllers may be deployed over the surveillance area following a network architecture. The task of sensors is to sense the smoke and report about fire occurrence to actuators and controllers [6] quickly. Then, the actuators and controllers may take further actions such as the actuators equipped with water sprinklers react for a fixed time.

The overall network performance of a CPS depends on the imposed communication protocols. Design of efficient communication protocols is one of the most prominent issues to enable optimized communication among devices. The communication protocols generally include Physical layer, Data Link layer, Network layer, Transport layer, and Application layer. However, design of Medium

Access Control (MAC) layer, Network layer, and Transport layer has been capturing attention of the researchers in literature. Design and modeling of these protocols while maintaining the quality of service (QoS) of network performance for CPSs are still in embryonic stage. To confirm reliable communication, these protocols should be designed considering the special features of CPSs such as device heterogeneity, dynamic nature of environment, and dynamic network topology. The functionality and capability of CPSs components such as sensors, actuators, and controllers vary according to the demand of applications. Hence, traditional MAC, Network, and Transport layer protocols of wireless sensor networks (WSNs) may not be well suited for communication over CPSs. For instance, the delay in transmitting data varies with different application requirements. In a fire handling application, quick delivery of data is needed. On the other hand, for an air-conditioning system that controls the temperature of a room, the transmission of data does not necessarily have to be quick [3]. While designing communication protocols, the transmission delay and reliability of data delivery process should be considered. Different devices incorporated in CPSs may have different requirements of data transmission.

As CPS integrate diverse devices and control the operation and actions among those devices, CPSs account for the necessity of efficient channel assignment. The heterogeneous nature of devices in CPSs demands multi-channel multi-radio communication in most of the cases. Another challenge is mobility management. Many CPSs such as vehicle management, efficient traffic control, and aircraft control include mobile devices. Mobility management in CPSs is totally different from traditional mobile system. CPSs integrate mobile devices with physical world which sometimes need human participation.

Cyber-physical systems typically demand communication over wireless medium. To connect multiple heterogeneous devices, efficient wireless channel utilization and energy efficiency over radio transmission are needed. In this aspect, efficient channel utilization can be achieved utilizing cognitive radio-based communication. Some CPSs integrate mobile devices such as smartphones with other devices such as laptop. For example, in a surveillance system, video footage is captured in real-time using sensors. Hence, a synchronized network is needed to control the integration among physical systems, humans, and the cyber space. Such network often requires scalability to maintain synchronization. To meet such requirement, cloud architecture is needed in CPS delivering computing powers. We can consider the communication for CPS as a chart mentioned in Fig. 3.2.

Here, we present several recent works pertinent to CPSs. We discussed about different layers of the architecture of CPS from different aspects of systems design. In Sect. 3.2, we describe the existing specialized protocols at different layers for CPSs such as MAC layer, Network layer, and Transport layer from the perspective of protocol design. Section 3.3 discusses about the issues of multi-radio communication for CPS. We provide the challenges and issues of mobility in CPS in Sect. 3.4. Section 3.5 outlines the efficient channel utilization of CPS using cognitive radio network. We illustrate the cloud architecture for CPS in Sect. 3.6. Finally, Sect. 3.7 indicates the future research issues and frontiers related to CPS.

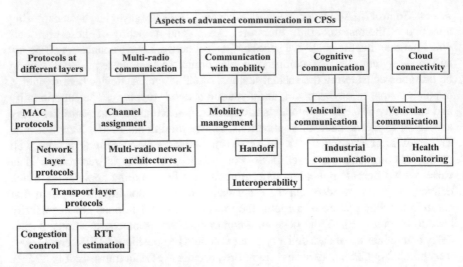

Fig. 3.2 Aspects of CPS communication

3.2 Specialized Protocols at Different Layers

Cyber-physical systems (CPSs) refer to the specialized form of embedded systems where computation, communication, and control are integrated with physical objects. CPSs are anticipated to be intelligently engineered systems having promising applications in diversified fields such as personal health care, medical services, intelligent transportation, scientific instruments, smart office, smart home, and public security. In CPSs, the interconnection of computing devices such as sensors and actuators is utilized in large-scale to complete multi-disciplinary tasks. Owing to the diversity of applications and computing devices, CPSs demand specialized communication protocols for computing devices to operate effectively and efficiently. Communication protocols generally consist of protocols of Physical layer, Data Link layer, Medium Access Control (MAC) layer, Network layer, Transport layer, and Application layer. Here, the impacts of protocols of Medium Access Control layer, Network layer, and Transport layer on network behavior of CPSs are worth of investigating in the literature. Hence, this section focuses on these protocols pertinent to CPSs.

3.2.1 Medium Access Control (MAC) Protocols

A suitable MAC layer protocol is required for interconnections among embedded devices in CPSs. IEEE 802.15.4 is the mostly investigated and widely adopted protocol in this regard [7]. IEEE 802.15.4 protocol is suitable protocol for short-range and low-power communication. Thus, IEEE 802.15.4 offers energy-efficient communication that often does not guarantee quality of services.

In recent years, low-rate wireless personal area networks (LR-WPAN) are being used extensively in many embedded applications. In these applications, IEEE 802.15.4 is being utilized as a wireless Medium Access Control (MAC) protocol. IEEE 802.15.4 has brought revolutionary emergence in LR-WPAN for its unique features [8]. IEEE 802.15.4 supports low data rate, low-power consumption, and low-cost wireless communication. The frequency band and other configuration of IEEE 802.15.4 are presented in Table 3.1. It also supports multi-hop network topology including star and peer-to-peer topology as shown in Figs. 3.3 and 3.4. In addition, IEEE 802.15.4 can operate on two modes, namely beacon-enabled and nonbeacon-enabled modes. In beacon-enabled mode, the device sends beacon frame periodically after beacon interval that can be configured [9]. IEEE 802.15.4 supports slotted Carrier Sense Multiple Access with Collision Avoidance (CSMA-CA) [8] protocol in beacon-enabled mode and unslotted CSMA-CA protocol in nonbeacon-

Table 3.1 Configuration of IEEE 802.15.4 [10, 11]

Property	Range
Raw data rate	868.0–868.6 MHz: 20 kb/s; 902–928 MHz: 40 kb/s; 2.4–2.483 GHz: 250 kb/s
Range	10–20 m
Latency	Down to 15 ms
Channels	868.0–868.6 MHz: one channel 902–928 MHz: up to 10 channels 2.4–2.483 GHz: up to 16 channels
Frequency band	Four PHYs: three for 868 MHz/915 MHz one for 2.4 GHz
Addressing	Short 8 bit or 64 bit IEEE
Channel access	CSMA-CA and slotted CSMA-CA
Temperature	Industrial temperature range −40 °C to +85 °C

Fig. 3.3 Star topology

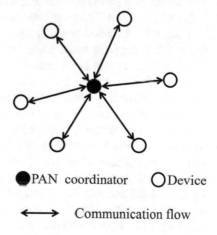

●PAN coordinator ○Device

⟷ Communication flow

Fig. 3.4 Peer-to-peer
architecture

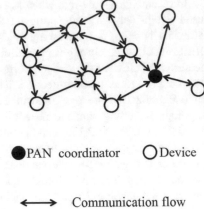

⬤ PAN coordinator   ◯ Device

⟷   Communication flow

enabled mode as the channel allocation mechanism. IEEE 802.15.4 is appropriate
for devices with limited resources. However, it is not well suited when data rate is
higher. Reason behind this incompatibility is that it does not exploit the notion of
Request to Send (RTS) and Clear to Send (CTS) to avoid collision.

The basic functionality of CPSs is based on wireless sensor and actuator
networks (WSANs) [4, 5]. WSANs refer to a generalized form of wireless sensor
network (WSNs), which includes deployment of sensors as well as actuators. IEEE
802.15.4 can be exploited in WSANs and hence in CPSs. One-hop star network
[7] considering deployment of all the nodes in each other's transmission range is
an example topology of such exploitation. The topology is presented in Fig. 3.5.
Here, both modes of 802.15.4 can be utilized. Irrespective of the adapted mode,
experimental results prove that default configuration of IEEE 802.15.4 cannot
provide best network performance for CPSs applications in various traffic load.
Default configuration refers to the frequency 2.4 GHz along with bit rate 250 Kbps.
The results also reveal the fact that it is very difficult to obtain a single generalized
IEEE 802.15.4 MAC configuration to ensure optimized network performance.

IEEE 802.11 is another MAC layer protocol, which is widely used in WSNs [12].
IEEE 802.11 supports two different access methods—distributed coordination function (DCF) and point coordination function (PCF). DCF is based on carrier sense
multiple access with collision avoidance (CSMA/CA) mechanism. The functionality
of CSMA protocol is that a device sense the medium before transmitting data. If the
medium is found to be busy, transmission is deferred. If the medium is found to
be free, the transmission takes place. IEEE 802.11 utilizes the notion of RTS/CTS
to avoid collision. Hence, IEEE 802.11 is suitable for long-range, high-bandwidth,
and high-power communication. For example, in environmental monitoring, sensors
monitor condition of the environment and then the sensed information is processed
by some local controllers. Afterwards, this resulting information is transmitted to
a central controller. The communication from local sensors to central controller
often demands high-bandwidth and high-data rate transmission. IEEE 802.11 can
be utilized for such communication.

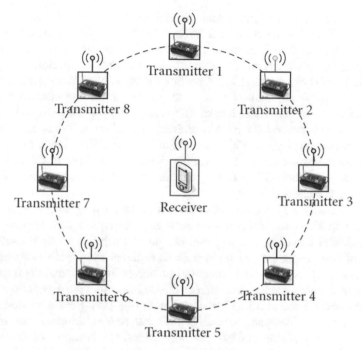

Fig. 3.5 Simulated star-network topology for cyber-physical systems [7]

3.2.2 Network Layer Protocols

Routing in Network layer is one of the most important aspects in data transmission over communication systems. CPSs demand efficient routing techniques to operate successfully. For efficient routing in CPSs, a number of challenges have to be addressed. For example, the network architecture itself varies for different applications of CPSs. This happens as devices such as sensors, actuators, and controllers are deployed at geographically different locations in CPSs. As a result, communication among these devices may be wired or wireless depending on the application requirements and position of deployment. Besides, these devices are not distributed in close proximity in many application scenarios. Hence, the communication may be single-hop or multi-hop. Moreover, mobility is yet another challenging issue, as some devices can be mobile and some can be static in a CPS.

In order to address the above-mentioned challenges, specialized routing protocols are needed for CPSs. Very few studies address the challenges in routing over CPSs to devise new specialized routing protocols. An example study [13] in this regard is based on IPv6 [14]. IPv6 is the most recent version of Internet protocol, which is an extension of IP version 4 (IPv4). Several structural updates in IP address have been made in IPv6 compared to that in IPv4. The most notable change is the use of 128 bits IP address in IPv6 instead of 32 bits used in IPv4, to support a large number of addressable nodes. Exploiting IPv6, multiple wireless sensor networks (WSNs) can be connected through the border routers [13]. The border routers operate based on IPv6 links. The main task of border routers is to translate IPv4 address to IPv6 address. Here, the overall network architecture can provide hop-by-hop forwarding and efficient routing along with the management of duty cycling. A deployment challenge for such architecture is the need to store 128 bits IP address and additional header information, which demands more storage compared to that needed for 32 bits address. Therefore, the drawback of using IPv6 is resource constraints, which is a major issue in CPSs. Nonetheless, as end-to-end communication between computing devices is required in CPSs, IP-based routing can be performed in CPSs. A network architecture [13] corresponding to such routing is presented in Fig. 3.6.

The requirements in routing techniques may vary owing to the requirements of specific applications of CPSs. One such requirements is data transmission to multiple destinations, i.e., multicasting. Multicast routing in CPSs is challenging because of two key issues which make CPSs different from traditional communication network. Firstly, the determination of uncertain destinations is required in CPSs. For example, in video streaming or file downloading, the destinations are the customers who are downloading the video clip or files [15]. Hence, the destinations are unknown and uncertain. Secondly, multicast routing demands the selection of optimal routing among multiple routing options. Conventional communication networks choose one set of paths for multicast routing which is called a routing mode. However, in CPSs, one routing mode may not provide best performance of system. Hence, multiple routing modes and switching among them is a challenging issue that needs special attention.

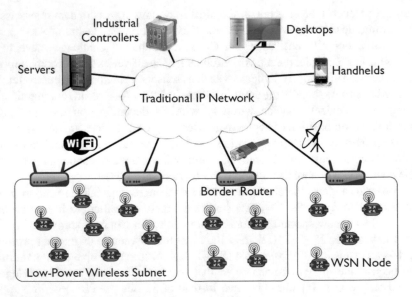

Fig. 3.6 Network architecture using border routers [13]

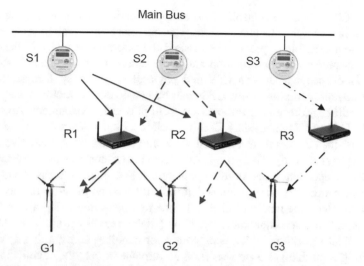

Fig. 3.7 Multi-cast routing in CPS [15]

An application of multicasting handling the above problems in CPSs is real-time voltage control in a smart grid [15]. Figure 3.7 exhibits an example scenario. Here, sensor, relay nodes, and distributed energy generators (controllers) are deployed in a grid that demands multicast routing. The sensors and controllers are integrated with communication interfaces assuming wired or wireless medium. Relay nodes are used as intermediate nodes to forward the observed information from sensors to controllers. Continuous data flow is considered from sensor to controller. Here, sensors monitor real-time voltage and send the sensed information to multiple controllers through relay nodes. Controllers evaluate real-time voltage and take necessary actions to stabilize the system. Besides, different routing modes are considered and dynamic switching among them is enabled. The main idea of proposed multicast routing is to determine a set of routing modes that can ensure stability of system dynamics and switch among these routing modes to achieve maximum system stability at a particular time. The connectivity of all sensors and controllers is stored in a matrix, which is called the feedback gain matrix. The connectivity is determined using bandwidth constraint. For each feasible connection between a sensor and a controller, the connection is added to the existing connections in the matrix, and thereby overall system stability is computed from the matrix periodically. For different connection scenarios, different states of matrix are determined and each state of connectivity matrix is considered as a routing mode. The routing mode with maximum system stability is selected for routing from sensors to controllers. The routing is demonstrated in Fig. 3.7 where one sensor forwards the sensed information to multiple relay nodes and hence to multiple controllers to deliver the observations accurately. Here, sensor $S1$ sends the data to relay nodes $R1$ and $R2$. $R1$ sends the data to controllers $G1$ and $G2$.

Many CPSs need the controllers to send information to sensors to take actions. For example, in a health care system, sensors sense information from physical environment (i.e., blood pressure) and send the information to controllers. Then, controllers process this information and send actions to the sensors. Thus, information is exchanged between CPS and physical environment forming a closed loop of actions. More precisely, CPSs can be considered as a closed loop system while sending and receiving information from physical environment. An example of study [16] of CPS considers CPS as a closed loop system. In this study, the whole distributed system is assumed as a combination of several dynamically formed subsystems [6]. The actions of these subsystems can be inter-dependent or totally independent based on the applications. The main concept of the work is that performance of whole system is determined based on the performance of each subsystem. Here, the nature of each subsystem is measured using a cost function which is called linear–quadratic regulator (LQR) cost function. LQR [17] is a mathematical algorithm, which is exploited for handling and running a controller controlling a machine or a process (i.e., an airplane, a vehicle, chemical reactor). LQR attempts to minimize a cost function with some weighting factors provided externally. The work [16] assume making one of weight factors of to depend on topology. The topology is determined regarding the whole communication network as an interconnection graph. Consequently, this topology-dependent LQR cost function is used to find the cost of each subsystem. The performance of overall system is then attained using cost of each subsystem. The routing and communication among devices are selected to improve the performance of the whole system. One major finding of this work is the fact that adding communication edges to the interconnection graph to find the topology sometimes may degrade the overall system performance.

In the work presented in [18], the authors propose an efficient event aggregation method utilizing proximity queries in a wireless sensor network. A framework termed as spatial and temporary processing (STP) is devised which reduces the cost for query registration by eliminating proximity events that are unnecessary. It also selects small number of aggregator nodes to send proximity alarms to the base node.

It is possible that, in many CPSs, underwater objects are used to perform certain tasks. Hence, under network communication architecture is an important issue in CPS. The study [19] proposes an energy saving tracking method which is based on local search for underwater wireless sensor network (UWSN). The main concept here is to keep active minimum number of sensors with a view to increasing network lifetime.

The research work done in [20] proposes an ant-colony meta-heuristics-based efficient collaborative routing mechanism. Here, best possible routing is constructed by making virtual circuits considering the load.

3.2.3   Transport Layer Protocols

CPSs consist of several embedded devices that are often deployed in an ad hoc manner over unpredictable environments. Therefore, communication among these devices may be single-hop or multi-hop pertaining to applications' requirements. As CPSs may be considered as wireless sensor and actuator networks (WSANs), they necessitate reliable and consistent data transmission to maintain robust communication among sensors and actuators. Hence, reliable Transport layer protocols are obligatory for CPSs. Existing Transport layer protocols may not be suitable for CPSs owing to their distinguished features, such as limited resource and low power. Problems that may generally arise in data transmission over CPSs can be categorized in two main aspects. First, packets may be dropped or lost owing to not having a suitable congestion control mechanism. Second, reliable data transmission may be interrupted due to not sustaining appropriate time synchronization (i.e., not having a good estimate of round trip time). Hence, this section presents these two aspects pertaining to CPSs in details.

3.2.4   Congestion Control Mechanisms

Achieving an optimized performance of Transport protocols over CPSs is always a challenging issue owing to the unique characteristics of CPSs. The notion of reliable data transmission in CPSs differs from that over other conventional networks. Here, performance of data transmission suffers for having a completely different type of a network with various embedded devices mostly connected in an ad hoc manner. Besides, the wireless medium in CPSs imposes lossy and non-deterministic data transmissions compared to mostly lossless and deterministic data transmission over wired networks. In addition, CPSs may contain different types of embedded devices resulting in heterogeneity. Hence, data transmission may have to adapt with both specialized low-bandwidth and high-bandwidth radios to meet requirements of the applications.

As CPSs often resemble WSANs, data transmission over CPSs frequently experiences propagation loss, multi-path routing, hidden station problems, etc. These problems rarely affect consecutive failures of data transmission attempts which is a considerable aspect for ensuring reliable data transmission. The main reason behind consecutive failures in data transmission is congestion in the underlined operating network. Congestion mainly occurs when two or more packets collide at the same time in a network. Congestion presents a classical obstacle to reliable data transmission.

A publish/subscribe-based middleware architecture, namely real-time data distribution service (RDDS) [21] enables reliable data dissemination over CPSs. This study utilizes the two traditional Transport protocols TCP and UDP for communication. Publish/subscribe architecture has gained popularity in recent distributed applications [22] because of its different features from conventional point-to-point architecture (i.e., client–server architecture). Here, producer (publisher) publishes events (sensed data on a topic of interest) and these events are broadcasted by server. Consequently, subscriber acquires this published data when needed while maintaining proper synchronization. Publish/subscribe architecture dynamically adds and removes publisher/subscriber which makes it a suitable communication system for a large-scale CPS. For example, in a search and rescue task during a fire accident in a building, the rescuing firefighters carry PDAs to gather data from nearby sensors to observe the dynamic condition of the building. Each firefighter's PDA can only acquire limited observations from nearby sensors. Hence, to get an overall information of the whole situation, all PDAs may have to combine the gathered real-time data by sharing synchronously [22]. These type of situation demand fusion of data from all sources (PDAs). Other examples of such CPSs are vehicular network, traffic control, and future combat systems.

The study [21] mainly focuses on reliable data transmission using two approaches. Firstly, for slow or unstable network, semantics-aware communication is exploited which refers to modeling of data streams utilizing lightweight physical models. In semantics-aware data stream modeling, same model is used for both publisher and corresponding subscribers to reduce computation and communication overhead. Secondly, to improve the quality of real-time data distribution and to achieve robustness, a reactive feedback mechanism at the publishers and the proactive feed-forward mechanisms at the subscribers are incorporated. Figure 3.8 shows a high-level architecture of RDDS. Here, each firefighter can participate as both publisher and subscriber to the sensor streams. T_{fire} is a topic of interest in which the generated events (data streams) are published by each entity (as a publisher) and each entity (as a subscriber) can also subscribe to the data collected from sensors. Quality-of-service (QoS)/quality-of-data (QoD) is maintained by a broker in a centralized manner. Transport layer protocols TCP and UDP are

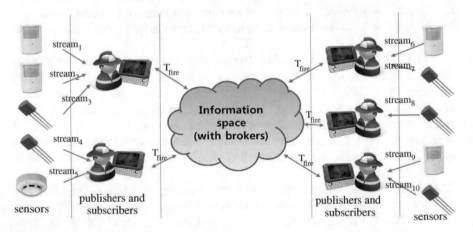

Fig. 3.8 Real-time data distribution service in cyber-physical system [21]

exploited for communication in lossy and unstable network. Here, packets are dropped following a random probability. The experiment results [21] reveal that UDP protocol is well suited for lossy communication as packet drop ratio with respect to exchanged number of messages remains stable while using UDP. On the other hand, while using TCP, the packet drop ratio increases with an increase in number of exchanged messages. This happens because UDP does not retransmit the lost messages, hence the total number of exchanged messages remains unchanged in lossy communication. However, while considering quality of data, TCP performs better than UDP. The quality of data degrades with an increase in communication load in UDP.

Maintaining importance of data is a crucial issue in congestion control for reliable data transmission over CPSs. Importance of data cannot be measured by assigning a static priority value or deadline [23]. It depends on dynamic and unpredictable state of the physical world. The sensing behavior of wireless sensor networks is widely distributed in many recent applications and hence, it demands accurate estimation of monitored physical measurements. For example, a CPS consisting of several wireless nodes along with base stations may be used to observe temperature and humidity distributions in a surveillance area [23]. The data can be sensed (sampled) with a fixed rate (i.e., once every second). Such type of sensing is called spatio-temporal as it provides a value in space and time. Transmitting this spatio-temporal data to base station using multi-hop routing sometimes results in congestion. Because of congestion, a node cannot transmit all the samples it observes to the next-hop node. Therefore, specialized congestion control mechanism is needed for collecting data with a view to maximizing the estimation accuracy of the data for such applications. This approach [23] considers different concurrent applications in the underlying CPS with different accuracy requirements. Here, relative importance of data is taken into account to minimize the overall estimation error. The data collected from different locations and at different times is summarized and utilized as a tool to control data transfer along with excluding congestion without increasing overall error in sensing the physical environment.

The main concept of this work [23] is that every node forwards its reading to corresponding base station regarding zero estimation error at starting. However, when congestion occurs, nodes experiencing congestion reduce their data transmission considering some estimation error to eliminate congestion. Hence, to eliminate congestion the minimal allowable estimation error is obtained. Here, each application is composed of several nodes along with a base station. Sensors sense data and send this data to base station. The connection among nodes is represented by a tree where the root of tree is the base station. Hence, sensed measurements are aggregated at intermediate nodes and are forwarded to parent node until it reaches to base station. Here, each application is assumed to accept a maximum tolerable error. The error is controlled locally in a neighborhood nodes to minimize the overall error. To determine the value of current error, the measurements at a node and the values accessible by its parent are compared. The data flow of a node is then controlled based on value of current error. The current error is kept less than or equal to the maximum tolerable error while controlling the output data flow. When congestion is detected, the value of maximum tolerable error is increased periodically until congestion is eliminated. Here, adaptive data summarization and aggregation are performed at intermediate nodes which ensure that the current error does not exceed the maximum tolerable error. This congestion control mechanism ensures more accurate estimation along with minimal communication overhead. However, for applications where the measurements of data need to be 100% accurate, this scheme cannot be applied because it allows some acceptable error to control congestion.

Determining the congestion window size is another important aspect while controlling congestion for reliable data transmission. Probability theory may be applied to do so [24]. Furthermore, an artificial intelligence-based congestion control technique [25] confirms the selection of optimal congestion window size over wireless mesh networks (WMNs) [26] in this regard. A wireless mesh network is a network where each node is connected to any other node. The distinguished characteristics of WMNs such as lossy and unpredictable environment in communication, data transmission without any base station, and similar pattern in traffic imposed by neighboring mesh nodes make reliable transmission in WMNs a challenging issue. These characteristics are also reasonable issues that may hinder reliable data transmission over CPSs. Here, neural networks (NNs) is used to control congestion which helps to omit the problems emerged from utilizing slow start, and congestion avoidance. NNs [27] refer to mathematical models that are used to represent the pattern of biological brains. NNs are composed of a number of neurons that unite with each other to execute some specific tasks. One or more inputs and a bias are processed by each neuron and thereby an output is generated based on the inputs. Figure 3.9 demonstrates the structure of a neuron where the output is a function of the sum of the weighted sum of the inputs and the bias. The proposed architecture utilizes multi-layer, feed-forward, zero bias NN with reinforcement learning to develop a congestion control mechanism. Multi-layer NN refers to a NN that have one input layer, one output layer, and multiple hidden layers. Reinforcement learning is a learning technique where the currently accessible inputs

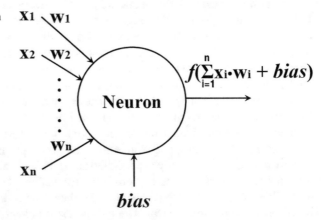

Fig. 3.9 A typical neuron in NN [25]

$$f\left(\sum_{i=1}^{n} x_i \cdot w_i + \textit{bias}\right)$$

are used to achieve an optimized output function. Feed-forward NN indicates a NN where direction of information is always forward, it means that the next output only depends on current inputs and independent of any intermediate outputs. Here, three parameters are selected as the inputs of NN and these parameters are utilized to obtain the output which is the optimal next congestion window size (cwnd). They are

- The number of consecutive timeouts,
- The number of duplicate acknowledgements (ACKs), and
- Current congestion window (cwnd) size.

The next cwnd size may increase, decrease, or remain fixed with respect to current cwnd size. The two inputs: the number of consecutive timeouts and number of duplicate ACKs are dynamically acquired from the performance of operating network. This congestion control mechanism is integrated with TCP which is called intelligent TCP (iTCP [28]).

The multi-layer NN is illustrated in Fig. 3.10 where one input layer, two hidden layers, and one output layer are shown. Three neurons are assumed as three individual inputs for consecutive timeouts (t_out), number of duplicate ACKs (dack), and current congestion window (cwnd) size. Then, the first hidden layer finds out the relative scale for increment, decrement, and no change of congestion window size. These three type of modifications represent three neurons (incr, decr, and same). Hence, these neurons calculate the relative weight of three types of update by taking inputs from each neuron in the input layer. These neurons also disseminate their outputs to the next hidden layer. The second hidden layer is used to find out the maximum order of update and the amount of that update. Here,

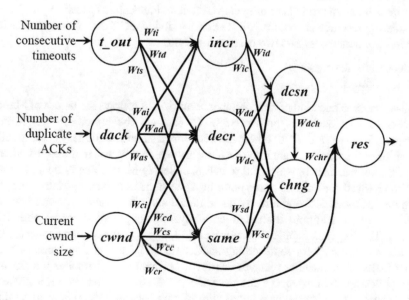

Fig. 3.10 Multi-layer structure of NN for determining the next congestion window size [25]

two neurons (decn and chng) are exploited to compute the desired two functions. The first neuron computes the maximum order of update which depends on the outputs forwarded by all neurons in the first hidden layer. Subsequently, the second neuron calculates the amount of update using current congestion window size and outputs of all the neurons from first hidden layer along with the maximum order of update from first neuron (decn). At last layer, only one neuron (res) is utilized to obtain the next congestion window size. The efficiency of this model depends on the appropriate regulations of weights and proper choice of the functions used in the neurons. The experimental evaluation of this scheme indicates that iTCP improves network performance in large coverage area of WMN with modest density.

3.2.5 Round Trip Time (RTT) Estimation Techniques

One of the most important parameters of timely and reliable data transmission is the estimation of round trip time accurately. Owing to special features of CPSs, a suitable round trip time estimation technique is essential to ensure reliable data transmission. Round trip time is the time delay between the transmitting of a packet and the receipt of its acknowledgement [29]. Estimating such round trip time ensures the successful and reliable delivery of data. If the acknowledgement of a packet is not received for too long, then the packet is considered to be lost and is retransmitted. Estimated round trip time is utilized to figure out when such retransmissions should take place. Accurate estimation of round trip time can remarkably improve the performance of underlying network.

A stateful round trip time estimation [30] scheme for wireless embedded systems utilizes an artificial intelligence (AI) technique called Q-learning [31] while considering resource constraint. This technique divides RTT estimation process into two independent cases. The two cases are as follows:

– Successful deliveries
– Packet drops.

Q-learning is one kind of reinforcement learning. It follows the nature of Markov process to take decisions optimally. In Q-learning, the effect of a decision is considered as a reward. After taking the decision, a state transition is occurred. Such transition to a particular state imposes a reward which is stored in a Reward-Matrix. Besides, another matrix is utilized which is called Q-Matrix that is updated based on the received reward and corresponding state transition. The Q-Matrix suggests optimum decision utilizing already taken decisions with a view to maximizing the next reward. In addition, it sometimes analyzes unexplored states to omit local optima. The key idea behind this approach [30, 32] is that two states (success and failure) are assumed for representing successful and failed transmission attempts individually. When sender receives acknowledgement then it implies a successful attempt and sender moves to a success state (S). Besides, when the retransmission timer expires, then it implies a failed attempt and forwards the sender to a failure

Fig. 3.11 State transition
diagram representing success
and failure of a wireless
transmission

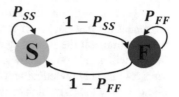

state (F). The state transition diagram representing success and failure is shown in
Fig. 3.11. Here, P_{xy} indicates the probability of a transition from state x to state y.
The probability of all transitions (P_{SS} and P_{FF}) is obtained using the corresponding
source states. Thus, such state transition diagram follows Markov process. Here,
exploiting this Markov process and Q-learning, the RTT in two different states is
estimated. This RTT estimation (QRTT) is integrated separately in the TCP. The
experimental results show that QRTT confirms significant improvement in network
performance for wireless embedded systems.

Adjusting round trip time by detecting sudden changes in traffic load in the
network is another approach for RTT estimation. The work [33] presents a RTT
estimation approach by detecting the traffic change for heterogeneous communication networks. The unique characteristics of heterogeneous communication
networks such as diversity in applications requirements, unpredictable condition
of traffic load, and combination of wired and wireless links make the reliable
data transmission of such networks a challenging issue. The traditional RTT
measurements are generally computed from packet acknowledgements. Considering
packet acknowledgements imposes delays which are caused by short-term traffic
changes in the network. Such short-lived traffic can be regarded as noise while
estimating RTT. The average RTT generally changes rapidly when the routing
path experiences a long-lived traffic flow. Hence, it demands specialized process
for filtering the short-lived (noise) and long-lived traffic load changes. It often
requires that the device should react to the quick changes and estimate RTT more
accurately to reduce packet loss and delays. However, first-order low-pass filter
which is used by conventional TCP cannot be used for quick changes as it only
depends on one parameter. A filter based on change detection can adapt for sudden
changes in the high traffic flow. The main idea of this work is introducing a filter
in RTT estimation depending on the change detection of traffic flow and detecting
only the long-lived traffic changes while considering short-lived changes as noise.
When network load increases depending on the application requirements, the RTT
should be accurately adjusted to handle the increased network load, traffic flows,
and sudden path changes. Here, an adaptive filter combining Kalman filter [34]
and CUSUM algorithm [35] is introduced which detects the long-lived changes.
Kalman filter [34] is a linear quadratic estimation algorithm which exploits a series
of measurements monitored over a time period, having noise and other inaccuracies.
It generates estimates of unknown variable with more accuracy. On the other hand,
CUSUM [35] algorithm is a one-sided cumulative sum that is utilized for observing
change detection. Here, Kalman filter and CUSUM algorithm provide an adaptive
and flexible filtering which achieves significantly better accuracy in RTT estimation.

In general, when a transmitter sends a packet, it waits to receive the acknowledgement for that packet. If it does not receive the acknowledgement for too long, it will retransmit the packet again. If the receiver receives the retransmitted packet, then it will send the acknowledgement. In traditional TCP, when the transmitter receives this acknowledgement, there is no way to determine which transmission is being acknowledged which causes a major problem called *retransmission ambiguity* [29]. A study called Karn's algorithm [29] addresses the retransmission ambiguity problem. The key idea of this algorithm is to avoid the RTT measurements for retransmitted packets for estimating RTT. Here, when transmitter receives an acknowledgement for a packet that has been retransmitted (sent more than once), it will ignore any round trip time for this packet. There exists another metric called *retransmission timeout* (RTO) which is the time period that a sender has to wait for a sent packet to be acknowledged [29]. Retransmission timeout solely depends on RTT and is computed using RTT. RTO is calculated as a function of RTT in the conventional TCP. However, RTO does not only depend on RTT, it also depends on some other metrics (i.e., congestion window size). The work done in [36] finds the optimal RTO using congestion window size and RTT with a view to maximizing the throughput of network. Here, the intuition that the larger the congestion window size, the longer the optimal RTO [36] is taken into account for determining the RTO. The optimal RTO is computed as a function of RTT and congestion window size. The optimal RTO maximizes the TCP throughput which has been proved by experimental results [36].

3.2.6   Cyber-Physical Systems and Internet of Things

The integration of embedded computing devices, human and physical environment constitute a cyber-physical systems (CPS) in which these entities are connected by a communication infrastructure. On the other hand, the Internet of Things (IoT) indicates to the interconnection of heterogeneous end-devices which communicate through Internet. These end-devices refer to sensors, actuator, RFID, embedded computer, laptops, mobile devices, smartphones, smart devices, etc. IoTs are envisioned to be a technology that enables decentralized control among the interconnected objects. These objects are capable of sensing, processing, storing, and networking. These objects can also act as intelligent agents and can share information with people and other devices which can be part of an interconnected CPS.

Therefore, IoT-enabled CPS demands special concern considering different communication issues. There exist many communication protocols which are used to connect the things to the Internet. IoT enables end-devices to be directly connected to the Internet utilizing cellular technologies such as 2G/3G and 5G [37, 38]. Besides, these devices can communicate through a gateway to the Internet [39].

When the devices connect to the Internet through a gateway, it forms a local area network. This type of connection generally refers to the machine to machine (M2M) network using different radios such as Zigbee [40] (use the IEEE 802.15.4 Standard), Wi-Fi (use the IEEE 802.11 Standard), Bluetooth (use the IEEE 802.15.1), and 6LowPAN [41] over Zigbee (use IPv6 over Low Power Personal Area Networks) [39]. Irrespective of type of the wireless communication that is used to establish M2M network, all the end-devices should be able to provide their information (data) to the Internet. This task can be executed by utilizing a web server or by deploying cloud. For M2M communication in IoT, there exist some standards effort such as 3gpp [42] or ETSI.

The Third Generation Partnership Project (3GPP) was established in 1998 to develop specifications for advanced mobile communications by the team European Telecommunications Standards Institute (ETSI) [43]. 3GPP standardization provides a recent NB (narrowband) radio technology to support the advanced requirements of the IoT. It will support a large number of devices with low-throughput, increase indoor coverage, low-power consumption, and optimized network architecture ensuring security, quality of service, and radio access which are the basic requirements for CPS.

Considering the communication protocols, protocols for the end-user Application layer is a major concern as the end-devices are heterogeneous. To address different requirements of communication, various protocols have been proposed by the researchers such as Advanced Message Queuing Protocol (AMQP) [44], Message Queue Telemetry Transport (MQTT) [45], and Extensible Messaging and Presence Protocol (XMPP) [39].

The Advanced Message Queuing Protocol (AMQP) is a protocol that has been developed on the basis of the financial industry. The mechanism of this protocol is that it can use various Transport protocols; however, it considers a reliable transport protocol like TCP as an underlying protocol [44]. Asynchronous publish or subscribe communication with messaging is supported by AMQP. It has store-and-forward feature which is the main benefit of AMQP. This feature confirms reliability in such a state when the network is disrupted [46]. AMQP confirms reliability using different message-delivery options. For addressing different needs and conditions of CPS, AMQP can be a potential protocol for communication.

Message Queue Telemetry Transport (MQTT) is another protocol based on M2M communication. It is also an asynchronous publish/subscribe protocol like AMQP. Publish/subscribe protocols confirm the network bandwidth decrement. In MQTT, a broker acts as a server that contains topics [45]. MQTT confirms reliability by supporting different options for maintaining QoS level [39]. MQTT ensures low overhead compared to other TCP-based Application layer protocols [47]. MQTT brokers require username/password authentication for ensuring security.

There are many IoT platforms to address different requirements in different scenarios such as Amazon Web Services IoT Platform (AWS), Google Cloud Platform, and Microsoft Azure IoT Hub [48].

AWS was the first to turn cloud computing into an asset for IoT in 2004. It is a scalable platform which can provide support for billions of devices as well as trillions of collaboration between them. Besides, Google Cloud Platform is the one of best IoT platforms supporting web-scale processing, analytics, and machine intelligence. It offers security in the form of "Google Grade." This IoT platform also has private global fiber network [48]. Another IoT platform, Microsoft is enabling Internet of Things through their cloud services. Like Amazon, Google, it also has some other beneficial services which include data analysis using machine learning.

Resource constraints is another prominent issue in IoT as well as CPS [49]. For efficient data management and analytics considering the limited resources, a widely used approach is fog computing. Fog computing is a hierarchical distributed architecture which is also known as edge computing. In fog computing, data, storage, computation, networking, and applications are distributed in an efficient way between data source and cloud. It extends the basic functionalities of cloud computing to edge network. Fog computing focuses on reducing the amount of data transmitted to the cloud for processing, analysis, and storage, thereby improving efficiency. Geo-distributed applications such as pipeline monitoring, wireless sensor networks to monitor the environment, mobile applications such as smart connected vehicle, connected rail, and large-scale distributed control systems such as smart grid, smart traffic light systems require efficient data management, analysis, knowledge of where data is computed and stored. These characteristics can be found within CPS and IoT. The main idea of fog computing is that the data processing is performed in a data hub rather than transmitting to the cloud for processing. Thus, it reduces the amount of data sent to the cloud for analysis. Hence, this type of computing can be a suitable solution for data management in CPS ensuring optimized resource utilization.

3.3   Specialized Multi-Radio Communication

The functionality of CPSs is totally different from sensor networks. CPSs typically offer sophisticated systems of heterogeneous embedded devices. On the other hand, in sensor networks, homogeneous sensors are generally deployed in large amount. Thus, in CPSs, a device may have to connect and coordinate with another kind of device. For example, a sensor node has to communicate with actuators, controllers, and other types of sensors. Here, functionalities and channel allocation techniques are not same for sensors, actuators, and controllers. This is exactly where the necessity for multi-radio multi-channel technology comes into the play for CPSs. Hence, this section presents several existing channel assignment techniques that are suitable for CPSs along with the crucial factors that affect the channel allocation algorithm from the perspective of multi-radio communication.

3.3.1 Channel Assignment Techniques

Channel assignment techniques for multi-radio communications are one of the most prominent concerns for ensuring optimized network performance. Channel assignment depends on various important issues. Some issues are presented below.

1. Connectivity: Connectivity is probably the most crucial issue for channel assignment algorithms in CPSs. Connectivity ensures data transmission among different nodes in a network. The importance of connectivity in a mesh network is illustrated in Fig. 3.12. Here, all the nodes are designed with two network interfaces utilizing four channels. The channel assignment technique in Fig. 3.12a confirms connectivity among all the nodes as the nodes form a connected component exploiting their available interfaces. However, if the allocated channel between *A–B* is changed from *Channel4* to *Channel1*, then *B* and *D* will not have any interface left to select a shared or common channel. As a result, the network is partitioned into two connected components as shown in Fig. 3.12b. Hence, this allocation of channel does not ensure connectivity over the whole network. In addition to connectivity, another important concern that needs to be investigated for data transmission is called *interference* [50].

2. Interference: When two nodes transmit data concurrently and their transmissions are sensed from a common position, their data transmissions get distorted at that position. This phenomena is called interference. Some efficient techniques have been developed to discover interference such as Protocol Model and Physical Model [51]. In Protocol Model, two transmission ranges are considered to detect interference—transmission range and interference range. On the other hand, Physical Model uses a threshold value at receiver for successful reception.

Fig. 3.12 Importance of channel allocation to ensure connectivity. (**a**) One connected component. (**b**) Two connected component [50]

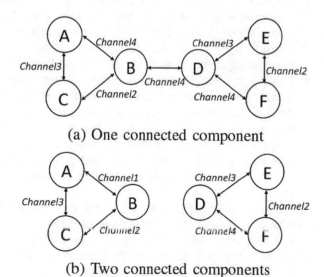

(a) One connected component

(b) Two connected components

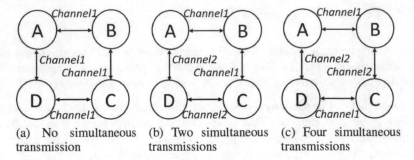

(a) No simultaneous transmission (b) Two simultaneous transmissions (c) Four simultaneous transmissions

Fig. 3.13 Illustration of channel diversity. (**a**) No simultaneous transmission. (**b**) Two simultaneous transmissions. (**c**) Four simultaneous transmissions [50]

The threshold value depends on the Signal to Interference and Noise Ratio (SINR). Physical Model is more applicable than Protocol Model in real scenario. However, the operation of Protocol Model is simpler than Physical Model. To accurately determine impact of an interference model, some important factors need to be considered such as path loss, signal reception, fading, and noise computation.

3. Channel Diversity: When all connected links of a node are exposed to non-overlapping channels, then this phenomena is called channel diversity. Channel diversity [50] is illustrated in Fig. 3.13. Figure 3.13a does not impose channel diversity as all the links are allocated to *Channel1*. However, Fig. 3.13b shows channel diversity as two links are assigned to *Channel2*, which enable two simultaneous transmissions for Node *A* and *C*. Finally, in Fig. 3.13c, all nodes can transmit exploiting two channels (*Channel1* and *Channel2*). Hence, channel assignment algorithm for CPSs should support channel diversity for allowing maximum number of simultaneous transmissions. A metric called *throughput* is considered to identify the efficiency of channel diversity. Throughput in a network is the average bit rate of transmissions. Most of the research studies focus on improving network performance by increasing throughput.

4. Dynamicity: Another important issue in channel assignment technique for CPSs is dynamicity. Dynamic nature of node activities and alivenesses in CPSs demands specialized channel assignment algorithm to cope up with any update in the network. Traffic condition, data flow, network topology, and physical environment are some parameters that can cause dynamic changes in the operation of CPSs. Hence, an efficient channel assignment algorithm should be designed for CPSs to make the whole network updated according to current status.

5. Distributiveness: The operation of CPSs can be controlled in centralized or distributed manner. Distributive channel assignment algorithms enable the embedded devices (nodes) to take own decisions of channel allocation. For efficient channel allocation, the devised algorithm should support distributiveness.

6. Mobility: Mobility is another important concern in CPSs. Mobile devices are exploited in CPSs depending on the application specifications. The channel

allocation algorithm should support efficient channel switching when a mobile device needs to switch channel because of the change in position to retain connectivity.

7. Fault Tolerance: CPSs may suffer from different types of faults such as fault in devices, link faults, and traffic congestion. Channel allocation algorithm should adapt to any kind of such faults through utilizing alternate channels to ensure connectivity.

There are some other criteria such as synchronization, scalability, stability, load balancing, utilization of fixed shared channel, and control overhead, which should be considered for efficient channel allocation algorithm for CPSs. There exist several channel assignment techniques that consider such criteria. A selected set of the techniques are presented below from the perspective of CPSs.

– Semidefinite programming (SDP)-based channel assignment approach [52] utilizes a centralized and static manner for channel allocation in multi-radio networks. The main idea behind this approach is that the channel is chosen randomly from k orthogonal channels while confirming minimum interference. Here, each device is equipped with multiple interfaces. Hence, a channel is allocated to both interfaces of each link to minimize interference conflicts. This technique ensures optimal channel assignment, though it is only applicable for orthogonal channels and does not consider traffic condition, external interference, etc. Nonetheless, simplicity and flexibility of this scheme make it an effective solution for data transmission in CPSs.

– Skeleton assisted partition FrEe (SAFE) [53] exploits randomized channel allocation in a distributive manner with a view to ensuring network connectivity over multi-hop communication. It considers two status of networks: the number of available channels and the number of available interfaces. SAFE allocates channel randomly if the number of usable channels is less than the double of accessible interfaces. When this condition is violated, network connectivity may be interrupted. In such a case, a connectivity graph called skeleton is formed and channels are reallocated to the edges to confirm connectivity. Here, the channel allocation randomly selects channel for all interfaces except one. That one interface is assigned as an edge of the skeleton. It also confirms channel allocation considering dynamic topology and increase in deployment. SAFE channel allocation is demonstrated in Fig. 3.14. In this scenario, each node has three available channels and two wireless interfaces. The channel allocation

Fig. 3.14 Illustration of channel allocation in SAFE channel allocation scheme [53]

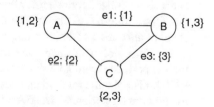

shown in Fig. 3.14 ensures no interference among links. However, this technique does not consider or impose any priority for the links.

- Adaptive dynamic channel allocation (ADCA) [54] is a dynamic channel allocation technique that operates over mesh topology. The technique considers two metrics of network performance for channel allocation—throughput and delay. Here, channel allocation is performed with a view to reducing delay and without diminishing throughput. These considerations make it a potential solution for channel allocation in CPSs having mesh topologies. In ADCA, each node maintains two interfaces—static and dynamic. The dynamic interface can switch channels. On the other hand, the static interfaces use fixed channels. ADCA supports maximization of throughput while allocating channels for static interfaces. Besides, dynamic interfaces choose channels by keeping queue for each neighbor. Priority of each neighbor is evaluated based on corresponding queue length and waiting time. Here, each node selects a channel in two steps. Firstly, each node selects a neighbor based on the priority, and secondly, channel negotiation is performed if the queue length is less than a specified threshold. This scheme supports negotiation of shared channel among more than two nodes at each time interval, which confirms reduced delay. Nonetheless, ADCA is not suitable for extreme traffic load, since the queue gets overloaded in the presence of extreme traffic load. In such cases, channel negotiation gets abandoned.
- Multi-radio breadth first search-based channel assignment (MRBFS-CA) [55] assigns channels over mesh topology. In this technique, a channel assignment server called CAS performs the channel allocation for the whole network periodically and informs all the nodes about the allocation. For channel assignment, CAS utilizes Protocol Model [51] to estimate interference assuming interference range as double of transmission range. It supports two different types of radios, namely default and non-default radios. This approach selects the default radio while ensuring minimization of interference between own network and external networks. In non-default radios, CAS constitutes a multi-radio conflict graph (MCG) [51] and applies BFS channel allocation over the MCG. To do so, both radios are differentiated according to the minimum hop counts from CAS and expected transmission time (ETT). It confirms connectivity along with minimizing interference. However, it causes high control overhead owing to broadcasting from the CAS and sending beacon messages from all nodes. The considered features of this technique also prevail in operations of CPSs. Hence, this scheme can be used for channel allocation in CPSs.
- Existing channel assignment techniques for multi-radio mobile ad hoc networks can also be utilized for channel allocation in multi-radio mobile CPSs. For example, Q-learning-based channel allocation [56] can be used for such CPSs. This channel allocation technique supports distributive and dynamic channel assignment. Here, nodes or agents are enabled to make decisions through analyzing their experience from an unknown environment enabling reinforcement learning [57]. It also obtains random and new operating points periodically going beyond the previous experience. The overall experience is maintained by a matrix called Q-matrix. Here, each decision taken by the agent is evaluated from its outcome

and the Q-matrix is updated accordingly. This technique was originally designed to ensure energy efficiency in sensor network and the decisions are evaluated in terms of energy efficiency. As it supports channel allocation with unknown characteristic of the environment, it can be utilized for assigning channels in CPSs. Here, the metrics of evaluation can be customized according to the CPS application requirements. Such metrics can be the number of transmissions, interference, connectivity, throughput, etc. The technique generally performs well for this flexible nature though it may not provide stable connectivity. Reason behind the instability of connectivity is that this decision making is performed at each individual node rather than performing any overall decision making for the whole network.

– Another approach for multi-channel allocation is channel assignment based on probability [58]. It offers a distributive and dynamic probabilistic channel usage-based channel allocation for wireless ad doc networks. This scheme keeps individual queues for each of the accessible channels. Here, channel is classified into two categories—fixed and switchable. Here, data reception is performed using fixed channels and data transmission is done using switchable channels. The persistence of channel allocation in fixed channel is longer than that in switchable channels. The underlying assumption of this technique makes it suitable for multi-radio channel allocation in CPSs applications having the capability of operating with fixed and switchable channels. In this technique, a node assigns channels randomly to all its interfaces and it updates the channel of fixed interfaces to a less utilized channel with some probability. This update occurs when the number of users on the common fixed channel grows larger. To keep this information, each node maintains a channel usage list and periodically updates it while exchanging the list with neighbor nodes. On the other hand, the switchable interfaces are assigned channels based on the oldest packet of the queue. The technique provides connectivity by using these switchable interfaces; however, it does not concede any interference cost while assigning the channels.

3.3.2   Multi-Radio Communication Architectures

CPSs demand specialized multi-radio multi-channel communication architecture to establish and maintain connectivity with the network. Most CPSs consist of heterogeneous devices. Moreover, network topology is also dynamic. This section describes some existing specialized multi-radio multi-channel architecture that can be utilized for many CPSs application for effective operation.

Energy efficiency is one of the most prominent issues in facilitating communication in CPSs. We know that radios of sensor nodes consume significant amount of energy. Since data transmission in CPSs is generally dynamic and unpredictable, hence data transmission can be frequent or infrequent based on application requirements. For infrequent data transmission, energy efficiency can be achieved by lowering the sleep-mode power consumption of multiple radios. In such scenarios, IEEE 802.15.4 is a potential solution since it supports low sleep-mode power consumptions. However, IEEE 802.15.4 is not appropriate for frequent data transmission. On the other hand, 802.11 supports frequent data transmission through its high bandwidth at the expense of more energy consumption. In addition, 802.11 can serve as a transmission-efficient radio since it consumes lower energy per bit of transmitted data than that of 802.15.4 radio. Nonetheless, 802.11 has high sleep-mode power. In order to handle the above-mentioned problems of both radios, specialized multi-radio techniques are needed that can combine the best qualities of both 802.11 and 802.15.4 radios.

In the study presented in [59, 60], an energy-efficient multi-radio architecture called Backpacking is proposed combining both short-range radio 802.15.4 and long-range radio 802.11. This study exploits energy efficiency of 802.11 radios from the perspective of data transmission and energy efficiency of 802.15.4 radios while remaining in non-active mode, even though both the types of radios work over the same frequency band (2.4 GHz ISM band).

The proposed architecture considers high data rate sensor networks. Here, 802.15.4 radio is utilized for transmitting sensed data from sensing nodes. The data sensing tasks are performed utilizing sensor nodes called originators. Originator nodes support only one 802.15.4 radio. After sensing the data by the originator nodes, data is accumulated and forwarded to the base station by another type of node called accumulator. These accumulator nodes are equipped with both 802.11 and 802.15.4 radios. Here, accumulator nodes receive or accumulate sensed data from originator nodes using 802.15.4 radios, and they forward these accumulated data to a base station using 802.11 radio. Hence, accumulated data gets backpacked using 802.11 radio. The network architecture considered in [59] is demonstrated in Fig. 3.15. Here, the hierarchy of collecting sensed data and sending the data to a base station is shown using 802.11 and 802.15.4 radios. There is an optimal deployment [59] density of the two radios for a network. A cross-layer mathematical estimation model[1] determines the density and provides a delicate balance of using both types of radios.

Another important concern in multi-radio communication is the heterogeneity of multiple radios. Heterogeneous multi-radios can improve the overall network performance in wireless networks. The characteristics of multiple homogeneous radios cannot always be adopted directly due to having different transmission ranges, bandwidths, and power consumptions. Thus, heterogeneous multiple radios come with some problems and challenges. The key design challenges of heterogeneous multiple radios are as follows [62]:

– Utilization and synchronization of multiple heterogeneous radios from a single device is a challenging task.

---

[1]A perfect modeling for such architectures is known to be infeasible [61].

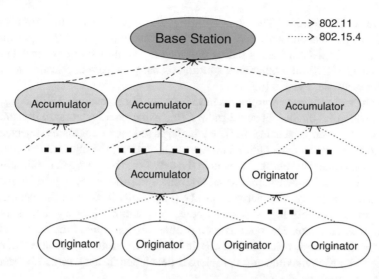

Fig. 3.15 Network architecture utilizing two different types of radios [59]

- For simultaneous utilization of multiple heterogeneous radios, the notion of data splitting comes into play. The optimal mechanism for efficient data splitting over multiple heterogeneous radios is another challenging task.
- How data splitting will be performed by the layers of the communication protocols is also another challenging task.

The above-mentioned challenges have been investigated in the literature [62]. It has been found that the radio with highest-bandwidth in the heterogeneous multiple radios performs well for low or moderate data rate multi-hop communication [62]. Nevertheless, this radio may show bad performance in high data rate applications. Considering these aspects, an approach [62] is proposed to perform concurrent activation of lower-bandwidth radios along with the best one to enhance network performance. Here, the activation of lower-bandwidth radio can improve the network performance in greater proportion than what the capability of lower-bandwidth radio adds with respect to the best one.

Another key issue is that the partitioning of data over multiple simultaneous radios demands optimized partitioning to achieve the best network performance. Here, optimal data splitting is determined taking into account some other factors such as network topology, data flow characteristics, and radio attributes. Furthermore, among multiple heterogeneous radios, the optimal point of data splitting changes according to the optimization parameters that can be configured by the application. Throughput, end-to-end delay, packet drop ratio are examples of such optimization parameters. Nonetheless, the optimal point of data splitting also depends on link quality and nature of a flow. Thus, the optimal data splitting over multiple heterogeneous radios can be achieved considering these aspects after selecting the optimization metrics from the application, which necessitates some user-level customization. This data splitting technique from application level is termed as simultaneous activation (SiAc) [62]. This approach of data partitioning among multiple heterogeneous radios provides a potential solution for data transmission in CPSs. Similar other data splitting techniques from application level also exist (for example, SymCo [63]).

The channel assignment problems have been jointly investigated with congestion control, which is suitable for multi-radio CPSs experiencing data rate transmission problems. In such a scheme [64], each deployed node regulates injected rate dynamically to achieve the optimal network utility. Here, the injected rate of a node is the rate at which packets are injected in the network. To achieve the optimality, the congestion control is considered as an optimization problem to maximize network performance or utility. The network utility [65] is obtained from the perspectives of metrics in multiple layers such as power consumption, delay, link data rate, and end-user data rate. The optimization problem is formulated mathematically by analyzing the network behavior and constraints for maximizing the network utility. The network utility maximization problem is developed through mixed-integer non-linear programming (MINLP).

3.4 Communication over Mobile Cyber-Physical Systems

Mobile cyber-physical system (MCPS) is a subcategory of CPSs in general [66]. The inherent distinguished feature of MCPS is mobility. Similar to the notion of CPSs where sensing the environment and reacting to sensed data are parts of processing and computation tasks, MCPS brings to the table non-stationary models of cyber-physical world. An example of a popular MCPS comprises smartphones [67]. Smartphones today are blessed with a physical dimension that allows them to be carried around in users' pockets. They have multi-core processing powers, considerable amount of data storage capability, multi-radio communication ability, and the support of high-level programming languages. However, not all MCPSs are blessed with such computing processes. Thus, for MCPSs with power consuming operations, tuning up variables and parameters such as network throughput, power consumption in radio operations, and data sensing and processing delays are very crucial.

MCPSs are different from conventional embedded mobile systems from the perspective that they are more human-centric demanding human participation and interaction [68]. For example, in MCPSs, human interventions process sensed data and make interpretations from it while the systems continue interacting with the physical world for us. Although the basic cycle of operations in MCPSs may appear to be similar, their applications are different. Moreover, the swarm of devices in an MCPS operating in a particular environment is different and heterogeneous compared to that operating in other networks [69, 70]. Consequently, this section discusses on MCPSs from three important aspects pertaining to mobility, namely mobility management, mobile handoff, and interoperability between carriers.

Before proceeding with discussions based on MCPSs, a general discussion regarding the state-of-the-art terms such as wireless mesh networks (WMNs), wireless sensor network (WSN), and mobile ad hoc networks (MANETs) is warranted. Although MCPSs, WMNs, WSNs, and MANETs are not necessarily completely different concepts, there still remain some important distinctions.

Ideally, the wireless ad hoc networks can be classified into the following three categories:

1. **Wireless mesh networks (WMNs):** The idea behind WMNs is to maintain a mesh topology over interconnected wireless nodes. IEEE 802.11 radio is mostly used because of its high-bandwidth and long-range networking capability. IEEE 802.11s is a new multi-hop networking technology specifically targeting mesh network topology. WMNs are self-organizing and self-configuring, which in turn reduces the setup time and maintenance cost. Because of low deployment cost, WMNs are preferred over single hop wired connection over a large area network. Moreover, because of alternative route to source–destination paths, the overall network reliability increases. WMNs also provide cross-domain interoperability via multi-point to multi-point architecture. Thus, interoperability among popular technologies such as Wi-Fi, WiMAX, Zigbee, cellular, and Bluetooth is possible [71].

2. **Wireless sensor network (WSN):** WSN is a network architecture of sensor nodes. Sensor nodes sense from environments, detect events, take actions immediately, or send data to a base station for interpretations. Some popular applications of WSNs are environment monitoring, industrial monitoring, smart home monitoring, etc.

3. **Mobile ad hoc network (MANET):** Wireless nodes are free to move. As a result, MANETs function over a dynamic topology with limited bandwidth constraints and variable link capacity. Some important applications and examples of MANET are battlefield communications, vehicle ad hoc network (VANET), internet-based mobile ad hoc network (iMANET), etc.

While WMNs, WSNs, and MANETs are not necessarily disjoint in core concepts, the recent trend of research and applications is mostly based on interactions between machines and physical world. In the following, we take a stride at understanding some differences between WSN, MANET, and MCPS [72].

– Routing capabilities and requirements are different for WSN, MANET, and MCPS. For example, MANET supports either unicast, multicast, or broadcast. On the other hand, WSN supports patterns such as query and response. In case of MCPS, several WSN may work together to form a system of interconnected sensor nodes. Cross-domain communication is also frequent in MCPS [73]. Such an example is the control of water gates of a dam through observing readings from several rain meters and water level measuring sensors [72]. In [74], a data collection approach is proposed for WSN considering mobile nodes. Here, instead of locating the exact mobile node, a node which is located in close proximity from exact node is selected to minimize the tour length.

– Mobility of nodes in case of MANET is arbitrary. Although the general norm of WSN is less mobility to stationary connected nodes, however, some controllable and uncontrollable mobility has also been studied [75]. In case of MCPS, it is well assumed that mobility is a requirement. The control state could either be human controlled or automated.

– Considering the routing capabilities and the possibility of having mobile nodes, MANETs have a random network formation. Contrary to the requirements of MANET, WSNs are more application and field specific. We already know that MCPS supports cross-domain communication. Hence, internet plays a dominant role in connecting cross-domain applications in MCPS.

3.4.1 Mobility Management

One of the key aspects of MCPS is research based on mobility types and models. Mobility modeling help researchers understand and define different aspects of mobility related problems. Since the network architecture of MCPS is closely related to mobile ad hoc networks which support microscopic view on mobility modeling, we resort to a similar view on MCPS. In this section, we will provide an overview of different types of cyber-physical systems, such as vehicle cyber-physical systems, airborne cyber-physical systems, and waterborne cyber-physical systems.

3.4.1.1 Airborne Cyber-Physical Systems

Airborne cyber-physical system (ACPS) is an old research topic. However, recent advancement in aviation and air warfare has led to massive investment in airborne cyber-physical systems. In ACPS, flight-paths, maneuver analytics and geometries, and multi-mode resources including ground-based nodes and control stations form the physical component. The cyber component consists of networking and communications, with computations and processing often off-loaded to cloud architecture [76, 77]. Figure 3.16 shows basic components and design principle of airborne cyber-physical systems.

An important component of ACPS is airborne network, or in other words, airborne MANET. It primarily consists of several subnets of nodes making connections with adjacent nodes while flying through a sea of virtual nodes [76]. Figure 3.17 shows an illustration of ACPS. It has few airborne and ground nodes. The requirements for ACPS are unique—mobility models should take into account smooth turns at high altitude, data transmission assurance, data authenticity, and data integrity. The purpose of mobility model is to provide a framework for studying connectivity, network performance, and decide which routing protocol performs optimally, given the constraints [78].

Fig. 3.16 Design principles of cyber-physical perspective of airborne network [76]

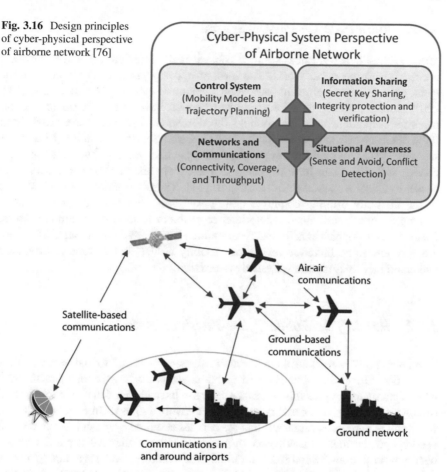

Fig. 3.17 Illustration of airborne network depicting cyber-physical systems [76]

Airborne objects tend to preserve motion in a straight line, that is, they maintain the same heading speed. In case they have to make a turn, it follows part of a large circular path [76]. In order to estimate path of an airborne vehicle, such as a jet fighter plane or a normal passenger plane, a mobility model should be able to capture such smooth turns. Some mobility models have been well rehearsed in literature. Random direction (RD) [79] and random waypoint model (RWD) [80] are two such models that theoretically come closer to airborne movements. In RWP model, the agent assumes a random destination. When it reaches to a particular destination, it pauses and then starts for the next destination. The non-uniform spatial distribution of nodes in RWP results in higher density towards the center and almost zero density towards the border region. Such a property is not desirable all the time in case of ACPS. Remedy to such non-uniformity in RWP is the random direction model. This model exhibits less fluctuation in node distribution. Above such theoretical studies, realistic models need to be developed in order to simulate and estimate ACPS.

3.4.1.2 Maritime Cyber-Physical Systems

The maritime cyber-physical systems consist of the interaction between physical processes and the cyber world. The idea of cyber-physical systems in maritime is to provide smart, efficient, and robust communication platform at sea [81]. Until recently, plenty of wireless infrastructures have been put up to support maritime communication. For example, the Wireless-Broadband-Access to Seaport (WISEPORT) in Singapore exploits WiMAX and fourth generation LTE in its sea area [82]. In maritime cyber-physical systems, vessels, buoys, ships, and coastal authorities can form a system of mesh network and exchange data from vessel-to-vessel or vessel-to-infrastructure [81]. There are ample research studies in the literature offering cognitive radio-based communication as an alternative to long distance multi-channel interference problem in maritime communication. Discussion on cognitive radio-based communication is available in Sect. 3.5. There are very few to no literature review on mobility models in maritime environment and conjointly linking cyber-physical systems with it.

3.4.2 Handoff in Mobile Cyber-Physical Systems

Handoff is an important attribute in mobile systems that need to maintain seamless connection to an end-point. Handoff or handover is the process of maintaining a user's active session even though he/she seems to have changed his/her current point of connection from one network to another network [83, 84]. In cellular network, when call is in progress, it means that the call has acquired a channel. When the call is complete, the channel is released. In-progress calls maintain a direct link with the nearby base stations, which forms a virtual area called cell as a region of coverage. When a mobile unit crosses over this cell into a new cell, it needs to switch its link to another base station and new frequency in order to continue [85].

Handoffs can be two types depending on how they were originally connected and how they are now connected to the new base station. Figure 3.18 shows two types of handoffs. Horizontal handoff supports the connectivity handover between two network base stations following similar network protocol stack [86]. For example, in case of cellular network, when a user communicates over his cell phone while moving in a car, horizontal handoff takes place. Vertical handoff takes place when the point of connectivity switches over between two networks supporting different network protocol stack. For example, a person switching over from cellular data to Wi-Fi while browsing internet [86]. The first step of handover is the initiation of the process, i.e., mobile node collects data regarding current link state, received signal strength (RSS), throughput, jitter, etc. These data help mobile device take a decision. Next, the mobile device executes the decision. If it has to execute handoff, then network authentication and authorization takes place to switch user's context from current state to a new state [83].

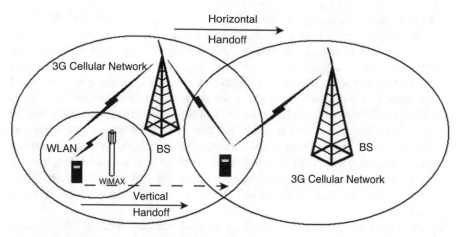

Fig. 3.18 Handoff in cellular network [83]

Currently, mobile handoff is mainly required for vehicular movements of mobile nodes. Such movements could be due to airborne vehicular movement, roadways vehicular movement, or high speed train movement. In case of high speed vehicular movement, there are three cases where problems regarding mobility persist [87]. First, frequent handover causes frequent re-selection of cells, hence communication quality will be disrupted. Second, Doppler frequency shift causes delay and failure in handover initiation and handover attempts, respectively. Finally, if the mobile node has a reasonably large mass or size, then it is probable that the mobile node will suffer from multipath fading. Hence, propagation modeling comes into the scenario.

3.4.3 Interoperability in Mobile Cyber-Physical Systems

The idea of interoperability is very crucial in terms of mobile cyber-physical systems. Current cyber-physical systems are designed and deployed to perform crucial tasks, for example, event-based environment monitoring, patient health care monitoring, industrial process automation, etc. Different devices are being used which constantly interact with physical world. They need to extract information from the interaction with physical world and get an interpretation from a computing or processing unit, or receive an action through human intervention. Despite the challenges prevalent in mobile sensing nodes, interoperability is yet another challenging domain. Interoperability demands inter-dependency and interconnection between different entities in an ecosystem of cyber-physical systems.

There are three important challenges in mobile cyber-physical systems interoperability [88].

1. **Heterogeneous sensor platforms:** The diversity in sensors and their platforms make programming those sensors complex and complicated. If we increase precision and granularity of sensor data reading, the effective bandwidth of the data extracted from the sensor increases. Moreover, since these sensors are mobile in nature, they are definitely battery based. Hence, power resource constraints is another concern while choosing sensing platforms. One way to solve heterogeneity in sensing systems is to provide abstraction to the gathered data. Essentially, some transformations may be applied to the raw stream of sensor data to produce a semantically meaningful stream. For example, instead of simply storing Wi-Fi RSSI data at each location for indoor localization, one can store corresponding access point location, their transmit power level, and their physical location (latitude–longitude). The latter can be termed as virtual sensor reading.

2. **Heterogeneous network:** While setting up a network, say WSN or WMN, a pool of commercial routers need to be bought and set up in our testbed. Commercial routers for Wi-Fi or other radios are less reliable in terms of fine tuning their transmit power level and received power level. Consider the case of Wi-Fi networks. Wi-Fi networks are susceptible to variable RSSI values in different conditions. With varying traffic load, the level of contentions and congestions will also be different. Among possible solutions to such dynamic problems, alternative measures to capturing data can be taken. For example, if the purpose of Wi-Fi network is to transmit data for localizing victims in disaster affected scenarios, a WSN can send audio data that semantically can explain the situation at that instant, such as loud noise, coughing, explosions, too loud, and too quiet for a long time.

3. **Heterogeneous applications:** Deployment of mobile cyber-physical systems is often preceded by extensive simulation and testing. For that, many software simulators are available to simulate events before considering real-life deployment. Such simulators are domain specific. Interoperability of heterogeneous applications can be reached by building multisimulation tool software. The idea is to reuse a simulation model for varying purposes.

One way of achieving network interoperability is to introduce a basic redundancy scheme [88]. In this system model, each device carries a cache to store and bundles for itself and others. When devices come within range of each other, they exchange these bundles. The goal of such a scheme is to introduce high reliability, short latency, and low storage cost. Such an approach can be implemented on a multinetwork topology. Links or nodes of different network access network topologies and their interactions. This is a multilevel approach to organize networks based on connectivity features and node stability.

3.5 Cognitive Radio-Based Communication over Cyber-Physical Systems

The Internet of Things (IoT) has ushered an era of connected devices. Starting from personal appliances and gadgets to the large-scale industrial applications, many devices are kept connected while fulfilling single to multi-objective tasks. Cyber-physical systems usually remain connected over wireless medium. Since multiple devices are in the play, the idea of efficient wireless channel utilization and energy efficiency over radio transmission are challenging. Hence, cognitive radio-based communication may be a solution to enhance efficient wireless channel utilization. Therefore, it can achieve energy efficiency and load balancing in CPS.

3.5.1 Cognitive Radio-Based Communication

The emergence of cognitive radio is from the idea of efficiently utilizing the radio electromagnetic spectrum. *Cognitive radio* is a software-defined radio built as an intelligent wireless communication system which learns from the environment and adapts to the changes in the wireless medium. This is a popular technology because of a couple of reasons [89–92]. Firstly, some frequency bands are partially utilized while some other bands are unused and unoccupied for the most of the time in wireless communication. Secondly, a part of the bands are more frequently utilized. Consequently, it is necessary to utilize the available spectrum judiciously, where unused and available spectrum are allocated to another user which is currently in need. According to [93], the primary objectives of cognitive radio can be summarized as follows:

– Communication with high reliability
– The efficient utilization of channels or radio spectrum.

We can see the basic steps of cognitive radio cycle in Fig. 3.19. Cognitive radio is a reality today. It is a combined effort of digital radio and computer software [94, 95]. Implementation of cognitive radio is based on some pre-defined tasks where signal processing techniques and machine learning tricks are being used. The study in [93] stresses upon three cognitive tasks—*radio scene analysis, transmitter power control and spectrum management, and channel-state estimation and modeling*. Figure 3.19 shows the steps. Here, the first and third tasks are carried by the receiver, while the second task is managed by the transmitter. The radio environment makes this interaction possible, and together they bind together and form a cognitive cycle [93]. Despite having a holistic overview of how cognition works in radio environment, it is nevertheless the task of the designer to employ the degree of cognition in cognitive radio-based communications. For example, the designer may pick a fixed spectrum for communication and adapt this cognitive cycle to that spectrum. Hence, both transmitter and receiver judiciously use the fixed

Fig. 3.19 Basic cognitive radio processes [93]

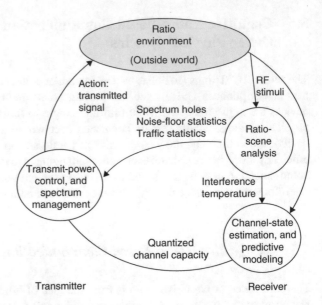

radio spectrum. It also designs this cognitive cycle across multiple band spectrum and reaches a target optimum that matches this designed performance and meets the expectations.

Pertaining to a macroscopic point of view of cognitive radio, there are two broader sets of cyber-physical systems. Firstly, *vehicular cyber-physical systems*. Secondly, *industrial cyber-physical systems*.

3.5.2 Cognitive Radio-Based Communication over Vehicular Cyber-Physical Systems

Vehicular cyber-physical systems belong to CPS. Here, vehicular and road networks are physical systems and computing and communication can be called cyber systems [96]. Vehicular cyber-physical systems have been emerged with different applications, such as road safety, green transportation, artificial intelligence assisted driving, and self-driving or automated driving. Figure 3.20 shows interaction of different components that is a part of vehicular cyber-physical systems.

Challenges Vehicular cyber-physical systems come with some problems and challenges [96]. Topology of vehicular networks changes constantly based on vehicular speed and mobility. Therefore, vehicles may need adaptive transmission power over wireless medium to establish reliable connectivity. The 5.9 GHz IEEE 802.11p Standard has been dedicated for vehicular communications. It has seven channels. These channels could be over-crowded in the presence of high density vehicular networks. A number of studies have also shown that the statically allocated wireless

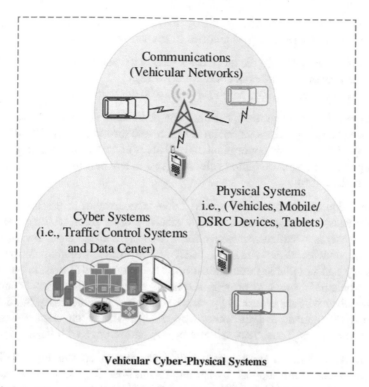

Fig. 3.20 Interaction of different system components in vehicular CPS [96]

Fig. 3.21 Cognitive cycle for vehicular cyber-physical systems [96]

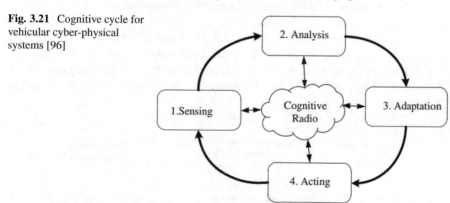

channels remain underutilized or idle most of the time. Moreover, the vehicular ad hoc network is susceptible to high dynamic and frequent changes of network topology. The high mobility of vehicles gives rise to several challenges also. As a result, vehicles need to adapt to network and communication parameters on the fly (Fig. 3.21).

The main idea of cognitive radio is to make efficient use of underutilized spectrum bands. There are two categories of users involved in this case.

- *The primary users (PUs)* are those users having a licensed spectrum band for use (such as network operators in cellular networks).
- *The secondary users (SUs)* are those users who are unlicensed.

However, they can use the spectrum provided that the PUs are absent and not using the spectrum (such as vehicular users in cellular band) [96]. The unlicensed *SU*s access idle channels opportunistically. This is done through sensing, analysis, and adaptation in cognitive radio cycle. As a result, any harmful interference to the *PU*s is avoided [96].

Transportation Cyber-Physical Systems There is an array of research that targeted spectrum access in cognitive radio networks. A similar track to vehicular cyber-physical systems is called *transportation cyber-physical systems*. Owing to high density of vehicles at any time, the IEEE 802.11p-based communication suffers from delay and unreliable communication. The study in [97] presents a solution to this problem. In order to provide a reliable communication, they assume that one transceiver will always query spectrum database for connectivity by remaining connected to internet, and the other transceiver will switch channels to adapt transmit parameters. Thus, there will not be any interference with PUs.

Figure 3.22 shows a system model diagram that is considered in [97]. The underlying architecture supports a distributed cloud-based system. The assumption is that, as soon as a vehicle starts for a destination, one of the transceivers (or GPS) suggests the best route to the destination. At the same time, this transceiver calculates the spectrum opportunities along the route towards destination, and also recommends for use en route. Since the vehicle itself is a *SU* cognitive radio, it periodically searches for spectrum opportunities in order to avoid interference with *PU*. For example, in Fig. 3.22, when the vehicle enters the road segment S2, it cannot use Wi-Fi channels 1 and 6 because residential Wi-Fi users (considered as PUs) are using them.

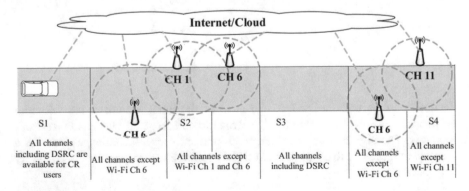

Fig. 3.22 System model for [97] where road-side Wi-Fi users are PUs

Fig. 3.23 Illustration of
platoon-based vehicular
cyber-physical systems [101]

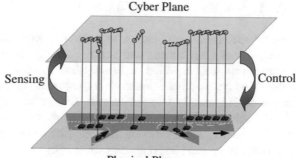

Impact for Mobility Reducing loss of data or communication during spectrum
mobility is an important research problem. Since the main tasks of cognitive
radio are spectrum sensing, management, mobility, and sharing [98]. Therefore,
cognitive radio is expected to solve spectrum mobility problems [99]. The work
in [100] introduces a fear inspired spectrum mobility scheme. Based on a survey
on different GSM service providers, a probabilistic deterministic finite automaton
can be proposed. The idea is to use fuzzy logic to represent different emotion states.
These emotion states quantitatively represent different communication parameters
and resemble the need for spectrum handoff (Fig. 3.23).

3.5.2.1   Clustered Vehicular Cyber-Physical System

While most of the work related to vehicular cyber-physical systems are based on
individual vehicles, some other work has opened a new paradigm of *clustered* or
platoon-based vehicular cyber-physical systems. In order to tackle issues related
to traffic congestion, mobility, and traffic dynamics, platoon-based driving pattern
has been suggested as a viable approach [102]. Platoon-based driving pattern has
several benefits. For example, vehicles in the same platoon will remain closer and
avoid congestion, streamlining vehicles one after another in a platoon will reduce
drag and thereby save energy consumption, and relatively fixed position among
the vehicles allows them to share data and communication channel and thereby
improves vehicular networking [103]. In order to support platoon-based cyber-
physical systems, two issues are involved:

– Issues related to vehicular networking and architecture
– Vehicle mobility models and traffic flow distribution.

[101]. The work done in [104] discusses about specific issues with vehicular mobility and handoff management. The challenges of vehicular communication caused
by high mobility and there some suggested solutions for host-based applications.

3.5.2.2   Maritime Cyber-Physical Systems

Limited spectrum opportunities, long distance communications, obstructions due to high density sea clutters and vessels, etc., are some of the problems of radio communication in maritime cyber-physical systems. Yang et al. [81] proposed a cognitive cooperative framework for providing opportunistic spectrum channel access in the sea. In this framework, SUs utilize the full transmission power for a specific period of time as a reward due to cooperation with PUs. Under this framework, each entity in the sea (for example, vessel, sea farm, oil/gas platform, etc.) should be equipped with sensing and communication devices. While registered or licensed users (or PUs) have spectrum opportunities, this framework makes opportunities for secondary users or unlicensed users to transmit messages on licensed spectrum. The framework proposed a game-theory-based resource allocation strategy which has been implemented in the MAC layer.

3.5.3   Cognitive Radio-Based Communication over Industrial Cyber-Physical Systems

Industrial cyber-physical systems are integrated systems that utilize computation power to control, influence, and interconnect physical systems. By the term physical systems, we abstract any physical body having mass and occupy a space around us. Consequently, in broader perspective, industrial cyber-physical systems mean the interaction between machine to machine and any other physical processes over cyber networks.

A broader range of research problems has been discussed in literature concerning industrial cyber-physical systems and cognitive radio networks. The most common topics concerning these domains are resource management in cognitive radio network, quality of service, state estimation, channel estimation, etc.

Resource Allocation The literature in [105] offers a comprehensive discussion on resource allocation problems in cognitive radio networks. This work has presented a systematic study on different design approaches such as signal-to-interference-noise-ratio, centralized or distributed framework based, etc. The work has also covered spectrum allocation and resource sharing options, in particular spectrum aggregation and frequency mobility. The work in [106] discusses about resource allocation problems based on some criteria, for example, interference, power, fairness, delay, topology nature—centralized or distributed, etc. Compared to this work, [105] provides basic mathematical formulation for resource allocation problems. They also discuss about the quality of service problems in relation to resource allocation problem. The literature in [107] discusses about cognitive radio network architecture on resource allocation. Their work is primarily based on efficient spectrum sensing and detection procedure in order to make proper resource allocation. The work in [81] considers resource allocation problems in maritime network topology and developed an opportunistic channel access framework for SUs over PUs.

The work in [108] discusses a multi-objective framework for resource allocation in cognitive radio network. Their design approach considers minimizing total transmit power, efficiency in energy harvesting, and minimizing interference power leakage-to-transmit power ratio. Their study revealed some interesting observations. For example, while allocating resources, the policy of minimizing total transmit power leads to low interference power leakage in general. Moreover, it has been found that energy harvesting maximization conflicts with the objective of minimizing transmit power.

Industrial cyber-physical systems often depend on sensor output to optimize industrial flow controls and control actuators and other peripheral systems. State estimation is the process of ensuring real-time monitoring of industrial processes and actions. Therefore, it is very important that sensor data reaches through wireless medium to the control node for further actions. This is a vital step for controlling system performance in industrial cyber-physical systems using integrated techniques of control and communication. As a result, much of the success of state estimation depends on reliable communication through wireless medium. In industrial wireless techniques, redundant channels are thus reserved to ensure reliability of wireless communication. However, introducing redundancy burdens the over-crowded ISM band. Cognitive radio communication has been emerged as one of the solutions to such problems. Cognitive radio can intelligently sense available spectrum and let SUs use unutilized spectrum from PUs. Different array of research methodologies have been adopted in this regard. The work in [109] proposed a cognitive radio enabled energy-efficiency maximization problem for state estimation convergence with the constraints of resource allocation. The given problem is a non-convex problem. They adopted a couple of relaxation techniques to transform the problem from non-convex problem to a convex problem. The work in [110] takes a different approach by modeling a channel sensing and switching mechanism called CHANCE. Their algorithm and process depends on channel quality and sensing accuracy, and took an iterative design approach by at first establishing a working technique for a single licensed and unlicensed channel. Then they extended their technique for multi-channel scenario. While state-of-the-art techniques rely heavily on network throughput, the work in [110] argues about considering communication reliability and state estimation performance as important variables. Nevertheless, a different category of research work considers two important parameters related to spectrum sensing—probability of detecting an unutilized spectrum and probability of false alarm generated by SU [111]. When the probability of spectrum detection is higher, PUs are protected since interference will be less to none. From the perspective of SUs, their objective is to lower the probability of false alarm generation. This in turn improves channel re-utilization and efficiency. The bottom line is that SUs always try to maximize their network throughput. The work in [111] mathematically formulates the sensing-throughput trade-off. Their work has revealed that an optimal sensing time can significantly improve throughput in SU. Finally, a recent trend in research has started focusing on blending of cognitive radios with multi-radio technologies in road to enhancing the performance metrics [112]. The studies suggest that such blending may not necessarily improve all the metrics.

3.6 Cloud-Connected Cyber-Physical Systems

In earlier days, classification of computing systems consisted of mainly two categories—traditional mainframe and desktop computers, and computing systems for controlling physical devices. Unlike the pasts, today's systems are interconnected. In other words, the physical and human systems are now connected through what we call to be a cyber space. Today, almost everyone has a desktop or laptop computer. Millions are the users of smartphones. These smartphones and laptops are now connected and synchronized. Surveillance systems today update video footage and camera positions instantly in real-time. From motion-aware systems such as airplanes or road vehicles to systems under ocean, there are constant interactions between physical systems, humans, and the cyber space. Combined together, there is a vast network of computing power and resources in sensors, actuators, and other networking devices. Such state of interconnected devices often requires scalability in terms of users and processing powers. Cloud architecture evolved to meet such needs, delivering computing powers and processors whenever needed. As such, systems and devices are continually appearing to us as ubiquitous and pervasive through exploiting cloud computing architectures.

3.6.1 Cloud Computing

Cloud computing and cloud services appeared to have a lasting impact in the ICT industry. The US National Institute of Standard and Technology upholds some key elements of cloud computing while defining cloud computing as follows [113]:

> Cloud computing is a model for enabling convenient, on demand network access to a shared pool of configurable computing resources (e.g., networks, servers, storage, applications, and services) that can be rapidly provisioned and released with minimal management effort or service provider interaction.

Over the time, several important service paradigms of cloud services evolved. Nowadays, there exist three prominent service models that collectively refer to cloud services [114]. The service models are

1. Software as a service (SaaS): Clouds offer software support to consumers. Cloud service providers or application developers release their software suit and deploy in the cloud to achieve scalability, speed, security, availability, and other resource optimizations. Consumers extract benefits of the cloud-powered applications

through abstraction models developed and provided by the service provider. Some examples of SaaS are email services (Yahoo mail, Gmail, Outlook, etc.), Google Docs, Overleaf, etc.

2. Platform as a service (PaaS): Clouds offer platforms to support the entire life cycle of software and services. Here, customers or developers can associate the entire development and deployment life cycle of their services with the cloud platform. Consequently, developers need not have to shift their development environment while prototyping their deployment cycle. Some popular PaaS are Amazon EC2, Google AppEngine, Microsoft Azure, etc.

3. Infrastructure as a service (IaaS): Clouds provide an abstraction to the consumer or developer of a huge computing resource under the hood. IaaS providers share access to virtually infinite amount of resources, such as devices, processing units, and storages. Virtualization is an important part of IaaS. The underlying idea is to set up virtual machines that are independent from the hardware and other similar virtual machines. Popular examples of IaaS are Amazon AWS, Microsoft Azure, IBM SmartCloud Enterprise, etc.

It is evident from the examples presented in SaaS, PaaS, and IaaS that cloud computing has surpassed a long path from sharing multi-core resources to sharing virtual environments ranging over mainframe computers to tiny wrist-band watches. Within such domains, two important sub-domains are highly related to cyber-physical systems, namely vehicular systems and health monitoring systems.

3.6.2 Cloud-Connected Vehicular Cyber-Physical Systems

In Sect. 3.5.2, we discussed about how cognitive radio-based communication solves some of the problems inherent with vehicular cyber-physical systems. In this section, we restrict our discussion to cloud-connected vehicular models only.

Vehicular networking has been a well-established research problem [115]. In recent times, some novel applications evolved exploiting sensors and actuators that help in decision making and autonomous control of vehicles [116]. Essentially, the control of these two dimensions (vehicles and sensors) has given rise to vehicular cyber-physical systems (VCPS) [116, 117]. Until recently, the idea of mobile cloud computing has emerged with VCPS as a coherent solution to emerging networked VCPS problems.

Figure 3.24 shows a hierarchical model of VCPS [118]. Based on spatial regions, there are three different layers in VCPS. First, the micro layer is a combination of intelligent embedded systems, environment sense and control factors, and human factors. Second, the meso layer is mostly related to cluster-based vehicle movement including networked vehicle routing called VANET. Finally, the macro layer provides control, information, and all kinds of services to improve quality of service (QoS), network throughput, real-time traffic updates, etc., to the clients or customers.

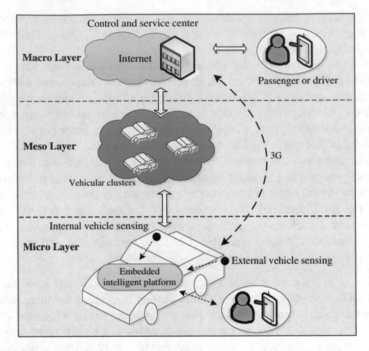

Fig. 3.24 Hierarchical model of VCPS [118]

The conceptual architecture of cloud-based VCPS [118] is primarily based on two basic ideas—first, mobile applications can be deployed to access larger and faster data storage centers for fast processing and information retrieval, and second, different mobile applications are developed based on differing architectural support to deliver cloud-based VCPS efficiently. In Fig. 3.25, the mobile devices take up the task of acting as a gateway to Internet connections outside the vehicle. A cloud server acts as a data storage and processing unit to take necessary decisions based on data gathered from several sensors and actuators in the vehicle through appropriate gateway mobile devices.

Contrary to explaining different architectural support, [119] takes on explaining state-of-the-art challenges on VCPS and cloud-connected support. Pertaining to existing VCPS problems with cloud-connectivity, the study in [119] explored the idea of context-aware cloud-connectivity where vehicular social networks and vehicular security have been explored. As a proof of concept, a context-aware dynamic car parking service has been proposed. To delineate on context-aware services, we can take some real-life scenarios as examples, such as real-time traffic live feed or availability of car parking facilities in a large shopping mall. The work in [120–122] provides some examples of context-aware traffic applications (Fig. 3.26).

Availability of parking services has been an intriguing problem among developed and developing nations. The usual scenario involves unavailability of parking spaces and parking-lot seekers wandering around for parking-lot availability. Some

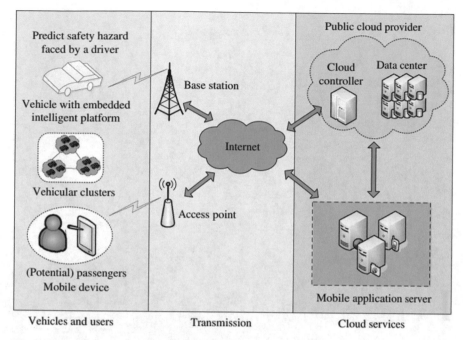

Fig. 3.25 Architectural model of cloud-based VCPS [118]

improved parking-lot facilities publish parking statuses over a billboard to inform incoming cars about possible parking spaces. Furthermore, a dynamic parking allocation scheme [119] may allow cars to park over a road temporarily, provided that it is not impeding any usual traffic movements. Contextually, the model employs road-traffic behavior. For example, it is well-known that traffic flow exceeds beyond capacity in some cities during rush hours such as morning and evening. Based on this contextual information, the system can update parking allocation schemes on wider and narrow roads, busy and non-busy roads, etc., accordingly. Moreover, in order to improve dynamicity, potentially empty parking lots are also considered within a window of specific time intervals. For example, the system can query a driver about his/her expected time of stay and departure from the parking facility. In this way, whoever is willing to park in the next few hours may consider parking in the same lot by observing potential empty time slots in future. As a result, context-aware optimizations may help improve traffic situations in countries where parking spaces are scarce.

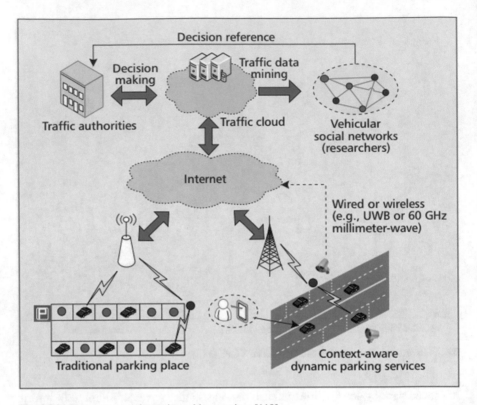

Fig. 3.26 Context-aware dynamic parking services [119]

3.6.3 Cloud-Connected Cyber-Physical Systems in Health Monitoring

Cloud-connected cyber-physical systems (CCPSs) combine the power of networked communication and computation with physical devices. Portability of such smart mobile devices, networking capabilities, and localization technologies has made CCPSs deliver promising tools for health care and monitoring [123]. One of the most popular mobile devices in this regard is the smartphones. Smartphones possess necessary computation power, multi-radio networking capabilities including GSM, Wi-Fi, Bluetooth, etc., localization capabilities utilizing GPS and other indoor-localization mechanisms based on Wi-Fi, inertial sensors, etc. [124]. Through off-loading complex computational steps, mobile CCPSs have opened an important field of research for patient monitoring, patient localization, and the health care services.

Existing studies based on CCPSs focus on quality of service (QoS), services related to general to specific illnesses, services provided to the elderlies, monitoring and localizing patients, etc. For example, several modern applications and devices have been developed to monitor different human vital signs [125]. Here, an array of devices generate huge amount of data at each period of time. Modern devices take help of cloud-powered applications to abstract and off-load huge computation required on the data from the data collecting devices to remote devices [126]. Thus, quality of service (QoS) becomes an important area of consideration for researchers and industries to ensure quality of data and quality of medical interventions in case of such off-loading. The area of research in this case often covers the trade-off among real-time collection and processing of data, prompt and swift health monitoring, dispatching timely medical interventions, quality of data sampling, etc.

Several key players [126] play important roles in this ecosystem. First, the physical aspect or the devices that measure and monitor data, second, the infrastructure that backs up this physical devices, for example, network architecture, processing and computing power, etc., and third, data analysis and management from the gathered data. Assuring quality of service among the three is a hard problem. Since health monitoring is a sensitive issue, it demands that the devices should offer near to 100% accuracy while accumulating data. While in some cases, these data are used for real-time analysis; however, in most cases, these data are used as a backbone machine learning analysis data to ensure future data are predicted with better accuracy. As a result, power consumption at device level shoots up considering the requirement of fine-grained sampling of personal health data. Increasing power consumption arrive at a price—for example, the fit bands that are available in our local market may consume more power if we want fine-grained sampling and monitoring of our health data. While ensuring data fidelity is desirable, it is equally less desirable to consume huge power in data collection only. Moreover, we know that, the more data that we collect for analysis, the better will be the outcome of the analysis. To ensure quantity of data, network infrastructure has to do a better job. However, networked components such as Wi-Fi, GSM, and Bluetooth consume data while transmitting data to the cloud. Consequently, data throttling rate has to reach an optimum where the trade-off between data quality and energy consumption in networked architecture is a research issue.

A recent trend of research is directed towards amalgamation of smart textile clothing and CCPS. The work in [127] introduces Wearable 2.0 for efficient health monitoring by exploiting human–cloud interaction. The idea is to ensure quality of experience (QoE) and QoS in smart clothing and advanced cloud services to deliver a reliable service to the customers. Here, they proposed a washable smart clothing. This smart clothing consists of sensors that continuously monitor health informatics data or human vital signs and then send these data to the analysis machine in the cloud. The proposed system harnesses the infinite power of cloud-based machines in analysis part. Figure 3.27 shows extension of the proposed idea to other domains where monitoring of human vital signs is essential.

Unlike the work in [127] where a generalized solution is provided, the work in [128] specifically engages elderly people in its design process. In this work, the authors monitored and observed how elderly people operate smart applications and appliances. In this aspect, the idea of energy consumption and energy efficiency has been explored. Elderly people often face difficulties in navigation and swift

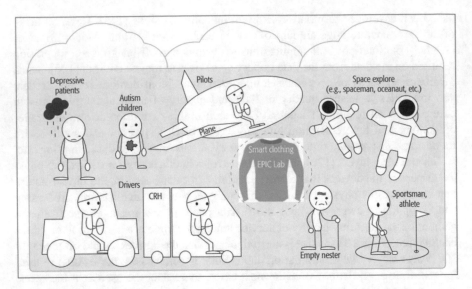

Fig. 3.27 Extending Wearable 2.0 to the cause of special groups of people and applications [127]

operation of electrical appliances. For example, elderly people may often fail to switch on/off electrical appliances. A cloud-based activity monitoring approach has been proposed in [128]. In this work, user's gesture or voice-based input is recognized to efficiently perform regular activities by the elderly people. Since high dimensional input is fed to the system, it is imperative that such input modalities are less likely to be processed by local processing units or embedded system units. As a result, a cloud-based approach is highly likely to suffice in this architecture.

3.7 Conclusion and Future Work

3.7.1 Future Work

There are still many research challenges involved in the cyber-physical systems communication. The following could be potential future research issues:

3.7.2 MAC Layer

- Self-configurable Mac protocols can be customized based on application requirements, traffic loads, and the environment.
- For collision free connection, simple Mac layer without RTS, CTS can be used for energy efficiency. When collision is detected, mac layer protocol with RTS,

CTS can be utilized. Mac layer protocols can be tuned with the link quality and capacity. Hybrid mac layer (SMAC, Mac 802.15.4, mac 802.11), Qos-aware Mac can be modified.

3.7.3 Network Layer

- Network layer considers platform heterogeneity (heterogeneity among sensors, actuators, and controllers).
- Routing protocols design affects the performance of CPS (considering energy efficiency, resource constraints, low latency, high throughput, low packet drop, low energy per bit, link quality, and path existence).
- Tree-based, cluster-based routing, and dynamic routing depend on the application requirement and architecture of CPS. Existing WSN routing protocols can be modified for CPS requirements also.

3.7.4 Transport Layer

- Dynamic congestion control mechanism can be explored based on the system architecture, packet drop, and window size. Optimal window size can be designed based on the requirements of CPS (i.e., traffic loads)
- Round trip time: Round trip time based on network performance (low packet drop, high packet delivery) can be tuned based on each subsystem of CPS. Existing UDP and TCP with specialized mechanism of RTT can be designed based on data importance, traffic load, network scale, and network behavior.

3.7.5 Multi-Channel Assignment

- Optimal multi-radio multi-channel assignment can be explored for each device in the CPS (based on connectivity, data importance, data flow, network topology, traffic load, and physical environment).
- Applications of existing WSN channel assignment techniques can be mapped for CPS based on system architecture and requirements.
- Power and resource management are most important issues for multi-radio multi-channel assignment. Optimal channel assignment minimizing these two metrics should be studied for different types of CPS applications.
- Multi-radio multi-channel assignment to minimize interference for different types of CPS applications (transportation system, health care system, and environmental monitoring).

– Channel assignment based on different metrics of different layer (cross-layer design) should be studied for CPS.

3.7.6   Mobile CPS

– There are many challenges involved in vertical and horizontal handover protocols for mobile nodes of CPS while communicating with a base station.
– Design automation of electric vehicles—electric vehicles are next generation vehicles that are power and battery charge critical. Electric vehicles may potentially face charging problems in regions where traffic mobility is slow and congested. For example, developing countries like Bangladesh, India, where energy constraints exist, induction charging or the energy harvesting mechanism for CPS can be explored.

3.7.7   Cognitive CPS

– Multi-objective optimization for cognitive radio used in CPS is a challenging issue.
– Transmit power minimization and energy harvesting efficiency maximization are conflicting design objectives in CPS.

3.7.8   Cloud CPS

– For complex industrial applications, different clients maneuver cloud-based CPS are being used based on their needs. A common framework for basic design goals and a prototype based on those design goals are yet to be established.
– Cloud-connected CPS, exchange information between different devices. This information exchange mechanism needs to be formalized based upon a common framework or protocol.
– Big data-based cloud CPS optimization is needed in terms of energy consumption, data fidelity, QoS, etc.
– Historical manufacturing process and performance can be integrated and maintained in a cloud-based knowledge repository. Combined with intelligent controllers, the future manufacturing processes can be improved continuously and design must be mapped for CPS.

CPS communication introduces many research areas, and the generated big volume of data has made this research area colorful and interesting.

References

1. T. Sanislav, L. Miclea, Cyber-physical systems-concept, challenges and research areas. J. Control Eng. Appl. Inf. **14**(2), 28–33 (2012)
2. S. Ali, S.B. Qaisar, H. Saeed, M.F. Khan, M. Naeem, A. Anpalagan, Network challenges for cyber physical systems with tiny wireless devices: a case study on reliable pipeline condition monitoring. Sensors **15**(4), 7172–7205 (2015)
3. N. Mohamed, J. Al-Jaroodi, S. Lazarova-Molnar, I. Jawhar, Middleware to support cyber-physical systems, in *2016 IEEE 35th International Performance Computing and Communications Conference (IPCCC)* (IEEE, Piscataway, 2016), pp. 1–3
4. J. Chen, X. Cao, P. Cheng, Y. Xiao, Y. Sun, Distributed collaborative control for industrial automation with wireless sensor and actuator networks. IEEE Trans. Ind. Electron. **57**(12), 4219–4230 (2010)
5. A. Nayak, I. Stojmenovic, *Wireless Sensor and Actuator Networks: Algorithms and Protocols for Scalable Coordination and Data Communication* (Wiley, Hoboken, 2010)
6. F. Xia, L. Ma, J. Dong, Y. Sun, Network QoS management in cyber-physical systems, in *International Conference on Embedded Software and Systems Symposia, 2008. ICESS Symposia'08* (IEEE, Piscataway, 2008), pp. 302–307
7. F. Xia, A. Vinel, R. Gao, L. Wang, T. Qiu, Evaluating IEEE 802.15. 4 for cyber-physical systems. EURASIP J. Wirel. Commun. Netw. **2011**(1), 596397 (2011)
8. L.S. Committee et al., *Part 15.4: Wireless Medium Access Control (mac) and Physical Layer (phy) Specifications for Low-rate Wireless Personal Area Networks (lr-wpans)* (IEEE Computer Society, Washington, 2003)
9. J. Zheng, M. Lee, A comprehensive performance study of IEEE 802.15. 4. Sensor Network Operations, 1–14 (2004)
10. E. Callaway, P. Gorday, L. Hester, J.A. Gutierrez, M. Naeve, B. Heile, V. Bahl, Home networking with IEEE 802.15. 4: a developing standard for low-rate wireless personal area networks. IEEE Commun. Mag. **40**(8), 70–77 (2002)
11. MAC, IEEE 802.15.4 (2003). https://en.wikipedia.org/wiki/IEEE_802.15.4. Accessed 20 Nov 2017
12. B.P. Crow, I. Widjaja, L. Kim, P.T. Sakai, IEEE 802.11 wireless local area networks. IEEE Commun. Mag. **35**(9), 116–126 (1997)
13. J.W. Hui, D.E. Culler, IP is dead, long live IP for wireless sensor networks, in *Proceedings of the 6th ACM Conference on Embedded Network Sensor Systems* (ACM, New York, 2008), pp. 15–28
14. S. Deering, R. Hinden, *RFC2460: Internet Protocol, Version 6 (IPv6) Specification* (RFC Editor, USA, 1998)
15. H. Li, L. Lai, H.V. Poor, Multicast routing for decentralized control of cyber physical systems with an application in smart grid. IEEE J. Sel. Areas Commun. **30**(6), 1097–1107 (2012)
16. C. Langbort, V. Gupta, Minimal interconnection topology in distributed control design. SIAM J. Control Optim. **48**(1), 397–413 (2009)
17. A. Bemporad, M. Morari, V. Dua, E.N. Pistikopoulos, The explicit linear quadratic regulator for constrained systems. Automatica **38**(1), 3–20 (2002)
18. M.R. Haque, M. Naznin, Monitoring cost reduction in sensor networks using proximity queries. JNW **6**(1), 4–11 (2011) [Online]. Available https://doi.org/10.4304/jnw.6.1.4-11
19. N. Majadi, M. Naznin, T. Ahmed, Energy efficient local search based target localization in an UWSN, in *12th IEEE International Conference on Wireless and Mobile Computing, Networking and Communications, WiMob 2016, New York, NY, October 17–19, 2016* (2016), pp. 1–8
20. M.S. Rahman, M. Naznin, T. Ahmed, Efficient routing in a sensor network using collaborative ants, in *Advances in Swarm Intelligence, 7th International Conference, ICSI 2016, Bali, June 25–30, 2016, Proceedings, Part II* (2016), pp. 333–340

21. W. Kang, K. Kapitanova, S.H. Son, Rdds: a real-time data distribution service for cyber-physical systems. IEEE Trans. Ind. Inf. **8**(2), 393–405 (2012)
22. S. Oh, J.-H. Kim, G. Fox, Real-time performance analysis for publish/subscribe systems. Futur. Gener. Comput. Syst. **26**(3), 318–323 (2010)
23. H. Ahmadi, T.F. Abdelzaher, I. Gupta, Congestion control for spatio-temporal data in cyber-physical systems, in *Proceedings of the 1st ACM/IEEE International Conference on Cyber-Physical Systems* (ACM, New York 2010), pp. 89–98
24. A.A. Al Islam, S.I. Alam, V. Raghunathan, S. Bagchi, Multi-armed bandit congestion control in multi-hop infrastructure wireless mesh networks, in *2012 IEEE 20th International Symposium on Modeling, Analysis & Simulation of Computer and Telecommunication Systems (MASCOTS)* (IEEE, Piscataway, 2012), pp. 31–40
25. A.A. Al Islam, V. Raghunathan, End-to-end congestion control in wireless mesh networks using a neural network, in *Wireless Communications and Networking Conference (WCNC), 2011 IEEE* (IEEE, Piscataway, 2011), pp. 677–682
26. I.F. Akyildiz, X. Wang, W. Wang, Wireless mesh networks: a survey. Comput. Netw. **47**(4), 445–487 (2005)
27. J.J. Hopfield, Neural networks and physical systems with emergent collective computational abilities. Proc. Natl. Acad. Sci. **79**(8), 2554–2558 (1982)
28. A.A. Al Islam, V. Raghunathan, iTCP: an intelligent TCP with neural network based end-to-end congestion control for ad-hoc multi-hop wireless mesh networks. Wirel. Netw. **21**(2), 581–610 (2015)
29. P. Karn, C. Partridge, Improving round-trip time estimates in reliable transport protocols. ACM SIGCOMM Comput. Commun. Rev. **17**(5), 2–7 (1987)
30. A.A. Al Islam, V. Raghunathan, $QRTT$: Stateful round trip time estimation for wireless embedded systems using Q-learning. IEEE Embed. Syst. Lett. **4**(4), 102–105 (2012)
31. P. Dayan, C. Watkins, Q-learning. Mach. Learn. **8**(3), 279–292 (1992)
32. A.A. Al Islam, V. Raghunathan, Evaluating Q-learning based stateful round trip time estimation over high-data-rate wireless sensor networks, in *2013 16th International Conference on Computer and Information Technology (ICCIT)* (IEEE, Piscataway, 2014), pp. 136–141
33. K. Jacobsson, H. Hjalmarsson, N. Möller, K.H. Johansson, Estimation of RTT and bandwidth for congestion control applications in communication networks, in *IEEE CDC, Paradise Island* (IEEE, Piscataway, 2004)
34. A.C. Harvey, *Forecasting, Structural Time Series Models and the Kalman Filter* (Cambridge University Press, Cambridge, 1990)
35. F. Gustafsson, F. Gustafsson, *Adaptive Filtering and Change Detection*, vol. 1 (Wiley, New York, 2000)
36. A. Kesselman, Y. Mansour, Optimizing TCP retransmission timeout, in *Networking-ICN 2005* (2005), pp. 133–140
37. L. Militano, G. Araniti, M. Condoluci, I. Farris, A. Iera, Device-to-device communications for 5g internet of things, in *IOT, EAI* (2015)
38. G. Camarillo, M.-A. Garcia-Martin, *The 3G IP Multimedia Subsystem (IMS): Merging the Internet and the Cellular Worlds* (Wiley, Chichester, 2007)
39. V. Karagiannis, P. Chatzimisios, F. Vazquez-Gallego, J. Alonso-Zarate, A survey on application layer protocols for the internet of things. Trans. IoT Cloud Comput. **3**(1), 11–17 (2015)
40. L. Fourati, L. Kamoun, Performance Analysis of IEEE 802.15.4/Zigbee Standard Under Real Time Constraints. International Journal of Computer Networks & Communications **315** (2011). https://doi.org/10.5121/ijcnc.2011.3517
41. Z. Shelby, C. Bormann, *6LoWPAN: The Wireless Embedded Internet*, vol. 43 (Wiley, Chichester, 2011)
42. J. Gozalvez, New 3gpp standard for iot [mobile radio]. IEEE Veh. Technol. Mag. **11**(1), 14–20 (2016)
43. Etsi 3gpp (1998). http://www.etsi.org/about/what-we-do/global-collaboration/3gpp. Accessed 20 Nov 2017

44. AMQP, Advanced message queuing protocol (2003). http://en.wikipedia.org/wiki/AdvancedMessageQueuingProtocol. Accessed 20 Nov 2017
45. S. Lee, H. Kim, D.-k. Hong, H. Ju, Correlation analysis of MQTT loss and delay according to QoS level, in *2013 International Conference on Information Networking (ICOIN)* (IEEE, Piscataway, 2013), pp. 714–717
46. F.T. Johnsen, T.H. Bloebaum, M. Avlesen, S. Spjelkavik, B. Vik, Evaluation of transport protocols for web services, in *Military Communications and Information Systems Conference (MCC), 2013* (IEEE, Piscataway, 2013), pp. 1–6
47. D. Thangavel, X. Ma, A. Valera, H.-X. Tan, C.K.-Y. Tan, Performance evaluation of MQTT and CoAP via a common middleware, in *2014 IEEE Ninth International Conference on Intelligent Sensors, Sensor Networks and Information Processing (ISSNIP)* (IEEE, Piscataway, 2014), pp. 1–6
48. IoT, Platform (2017). https://www.devteam.space/blog/10-best-internet-of-things-iot-cloud-platforms/. Accessed 20 Nov 2017
49. F. Bonomi, R. Milito, P. Natarajan, J. Zhu, Fog computing: a platform for internet of things and analytics, in *Big Data and Internet of Things: A Roadmap for Smart Environments* (Springer, Cham, 2014), pp. 169–186
50. A.A. Al Islam, M.J. Islam, N. Nurain, V. Raghunathan, Channel assignment techniques for multi-radio wireless mesh networks: a survey. IEEE Commun. Surv. Tutorials **18**(2), 988–1017 (2016)
51. P. Gupta, P.R. Kumar, The capacity of wireless networks. IEEE Trans. Inf. Theory **46**(2), 388–404 (2000)
52. H. Dinh, Y.-A. Kim, S. Lee, M. Shin, B. Wang, SDP-based approach for channel assignment in multi-radio wireless mesh networks. Networks **13**, 15 (2007)
53. M. Shin, S. Lee, Y.-a. Kim, Distributed channel assignment for multi-radio wireless networks, in *2006 IEEE International Conference on Mobile Adhoc and Sensor Systems (MASS)* (IEEE, Piscataway, 2006), pp. 417–426
54. Y. Ding, K. Pongaliur, L. Xiao, Channel allocation and routing in hybrid multichannel multiradio wireless mesh networks. IEEE Trans. Mobile Comput. **12**(2), 206–218 (2013)
55. K.N. Ramachandran, E.M. Belding-Royer, K.C. Almeroth, M.M. Buddhikot, Interference-aware channel assignment in multi-radio wireless mesh networks, in *Infocom*, vol. 6 (2006), pp. 1–12
56. J. Gummeson, D. Ganesan, M.D. Corner, P. Shenoy, An adaptive link layer for range diversity in multi-radio mobile sensor networks, in *INFOCOM 2009, IEEE* (IEEE, Piscataway, 2009), pp. 154–162
57. R.S. Sutton, A.G. Barto, *Introduction to Reinforcement Learning*, 2nd edn. (MIT Press, Cambridge, MA, USA, 2018). isbn:978-0-262-03924-6
58. P. Kyasanur, N.H. Vaidya, Routing and link-layer protocols for multi-channel multi-interface ad hoc wireless networks. ACM SIGMOBILE Mobile Comput. Commun. Rev. **10**(1), 31–43 (2006)
59. A.A. Al Islam, M.S. Hossain, V. Raghunathan, Y.C. Hu, Backpacking: deployment of heterogeneous radios in high data rate sensor networks, in *2011 Proceedings of 20th International Conference on Computer Communications and Networks (ICCCN)* (IEEE, Piscataway, 2011), pp. 1–8
60. A.A. Al Islam, M.S. Hossain, V. Raghunathan, Y.C. Hu, Backpacking: energy-efficient deployment of heterogeneous radios in multi-radio high-data-rate wireless sensor networks. IEEE Access **2**, 1281–1306 (2014)
61. A. Islam, V. Raghunathan, Assessing the viability of cross-layer modeling for asynchronous, multi-hop, ad-hoc wireless mesh networks, in *Proceedings of the 9th ACM International Symposium on Mobility Management and Wireless Access* (ACM, New York, 2011), pp. 147–152
62. A.A. Al Islam, V. Raghunathan, SiAc: simultaneous activation of heterogeneous radios in high data rate multi-hop wireless networks. Wirel. Netw. **21**(7), 2425–2452 (2015)

63. A.A. Al Islam, V. Raghunathan, Symco: symbiotic coexistence of single-hop and multi-hop transmissions in next-generation wireless mesh networks. Wirel. Netw. **21**(7), 2115–2136 (2015)
64. K. Li, Q. Liu, F. Wang, X. Xie, Joint optimal congestion control and channel assignment for multi-radio multi-channel wireless networks in cyber-physical systems, in *Symposia and Workshops on Ubiquitous, Autonomic and Trusted Computing, 2009. UIC-ATC'09* (IEEE, Piscataway, 2009), pp. 456–460
65. X. Lin, N.B. Shroff, R. Srikant, A tutorial on cross-layer optimization in wireless networks. IEEE J. Sel. Areas Commun. **24**(8), 1452–1463 (2006)
66. J. White, S. Clarke, C. Groba, B. Dougherty, C. Thompson, D.C. Schmidt, R&d challenges and solutions for mobile cyber-physical applications and supporting internet services. J. Internet Serv. Appl. **1**(1), 45–56 (2010)
67. M. Conti, S.K. Das, C. Bisdikian, M. Kumar, L.M. Ni, A. Passarella, G. Roussos, G. Tröster, G. Tsudik, F. Zambonelli, Looking ahead in pervasive computing: challenges and opportunities in the era of cyber–physical convergence. Pervasive Mob. Comput. **8**(1), 2–21 (2012)
68. X. Hu, T.H. Chu, H.C. Chan, V.C. Leung, Vita: a crowdsensing-oriented mobile cyber-physical system. IEEE Trans. Emerg. Top. Comput. **1**(1), 148–165 (2013)
69. R. Baheti, H. Gill, Cyber-physical systems, in *The Impact of Control Technology*, vol. 12 (2011), pp. 161–166
70. R.R. Rajkumar, I. Lee, L. Sha, J. Stankovic, Cyber-physical systems: the next computing revolution, in *Proceedings of the 47th Design Automation Conference* (ACM, New York, 2010), pp. 731–736
71. M. Seyedzadegan, M. Othman, B.M. Ali, S. Subramaniam, Wireless mesh networks: WMN overview, WMN architecture, in *International Conference on Communication Engineering and Networks IPCSIT*, vol. 19 (2011), p. 2
72. F.-J. Wu, Y.-F. Kao, Y.-C. Tseng, From wireless sensor networks towards cyber physical systems. Pervasive Mob. Comput. **7**(4), 397–413 (2011)
73. L. Han, S. Potter, G. Beckett, G. Pringle, S. Welch, S.-H. Koo, G. Wickler, A. Usmani, J.L. Torero, A. Tate, FireGrid: an e-infrastructure for next-generation emergency response support. J. Parallel Distrib. Comput. **70**(11), 1128–1141 (2010)
74. M.S. Rahman, M. Naznin, Shortening the tour-length of a mobile data collector in the WSN by the method of linear shortcut, in *Web Technologies and Applications – 15th Asia-Pacific Web Conference, APWeb 2013, Sydney, April 4–6, 2013. Proceedings* (2013), pp. 674–685
75. G. Xing, T. Wang, Z. Xie, W. Jia, Rendezvous planning in wireless sensor networks with mobile elements. IEEE Trans. Mob. Comput. **7**(12), 1430–1443 (2008)
76. K. Namuduri, Y. Wan, M. Gomathisankaran, R. Pendse, Airborne network: a cyber-physical system perspective, in *Proceedings of the first ACM MobiHoc Workshop on Airborne Networks and Communications* (ACM, New York, 2012), pp. 55–60
77. K. Sampigethaya, R. Poovendran, Aviation cyber–physical systems: foundations for future aircraft and air transport. Proc. IEEE **101**(8), 1834–1855 (2013)
78. G. Ravikiran, S. Singh, Influence of mobility models on the performance of routing protocols in ad-hoc wireless networks, in *2004 IEEE 59th Vehicular Technology Conference, 2004. VTC 2004-Spring*, vol. 4 (IEEE, Piscataway, 2004), pp. 2185–2189
79. P. Nain, D. Towsley, B. Liu, Z. Liu, Properties of random direction models, in *INFOCOM 2005. 24th Annual Joint Conference of the IEEE Computer and Communications Societies. Proceedings IEEE*, vol. 3 (IEEE, Piscataway, 2005), pp. 1897–1907
80. C. Bettstetter, H. Hartenstein, X. Pérez-Costa, Stochastic properties of the random waypoint mobility model. Wirel. Netw. **10**(5), 555–567 (2004)
81. T. Yang, H. Feng, C. Yang, Z. Sun, R. Deng, Resource allocation in cooperative cognitive radio networks towards secure maritime cyber physical systems, in *2016 IEEE Canadian Conference on Electrical and Computer Engineering (CCECE)* (IEEE, Piscataway, 2016), pp. 1–4
82. C. news, Maritime WiMAX network launched in Singapore (2008). http://www.cellular-news.com/story/29749.php. Accessed 24 June 2017

83. P.S. Jirapure, A.V. Vidhate, Survey and analysis of handoff decision strategies for heteroge-neous mobile wireless networks. Int. J. **4**(4), 703–713 (2014)

84. N. Nurain, T. Akter, H. Zannat, M.M. Akter, A.A. Al Islam, M.H. Kabir, General-purpose multi-objective vertical hand-off mechanism exploiting network dynamics, in *2015 IEEE 11th International Conference on Wireless and Mobile Computing, Networking and Communications (WiMob)* (IEEE, Piscataway, 2015), pp. 825–832

85. D. Hong, S.S. Rappaport, Traffic model and performance analysis for cellular mobile radio telephone systems with prioritized and nonprioritized handoff procedures. IEEE Trans. Veh. Technol. **35**(3), 77–92 (1986)

86. A. Ahmed, L.M. Boulahia, D. Gaiti, Enabling vertical handover decisions in heterogeneous wireless networks: a state-of-the-art and a classification. IEEE Commun. Surv. Tutorials **16**(2), 776–811 (2014)

87. Y. Zhou, B. Ai, Handover schemes and algorithms of high-speed mobile environment: a survey. Comput. Commun. **47**, 1–15 (2014)

88. K. Benson, C. Fracchia, G. Wang, Q. Zhu, S. Almomen, J. Cohn, L. D'arcy, D. Hoffman, M. Makai, J. Stamatakis et al., Scale: safe community awareness and alerting leveraging the internet of things. IEEE Commun. Mag. **53**(12), 27–34 (2015)

89. P. Kolodzy et al., Next generation communications: Kickoff meeting, in *Proc. DARPA*, vol. 10 (2001)

90. G. Staple, K. Werbach, The end of spectrum scarcity [spectrum allocation and utilization]. IEEE Spect. **41**(3), 48–52 (2004)

91. C.S. Hyder, A.A. Al Islam, L. Xiao, E. Torng, Interference aware reliable cooperative cognitive networks for real-time applications. IEEE Trans. Cogn. Commun. Netw. **2**(1), 53–67 (2016)

92. C.S. Hyder, A.A. Al Islam, L. Xiao, Enhancing reliability of real-time traffic via cooperative scheduling in cognitive radio networks, in *2015 IEEE 23rd International Symposium on Quality of Service (IWQoS)* (IEEE, Piscataway, 2015), pp. 249–254

93. S. Haykin, Cognitive radio: brain-empowered wireless communications. IEEE J. Sel. Areas Commun. **23**(2), 201–220 (2005)

94. M. Dillinger, K. Madani, N. Alonistioti, *Software Defined Radio: Architectures, Systems and Functions* (Wiley, Chichester, 2005)

95. J. Mitola, The software radio architecture. IEEE Communications Magazine **33**(5), 26–38 (1995)

96. D.B. Rawat, C. Bajracharya, Vehicular cyber physical systems (2016)

97. D.B. Rawat, S. Reddy, N. Sharma, B.B. Bista, S. Shetty, Cloud-assisted GPS-driven dynamic spectrum access in cognitive radio vehicular networks for transportation cyber physical systems, in *2015 IEEE Wireless Communications and Networking Conference (WCNC)* (IEEE, Piscataway, 2015), pp. 1942–1947

98. J. Mitola, Cognitive radio for flexible mobile multimedia communications, in *1999 IEEE International Workshop on Mobile Multimedia Communications, 1999 (MoMuC'99)* (IEEE, Piscataway, 1999), pp. 3–10

99. I.F. Akyildiz, W.-Y. Lee, M.C. Vuran, S. Mohanty, Next generation/dynamic spectrum access/cognitive radio wireless networks: a survey. Comput. Netw. **50**(13), 2127–2159 (2006)

100. F. Riaz, M.A. Niazi, Spectrum mobility in cognitive radio-based vehicular cyber physical networks: a fuzzy emotion-inspired scheme, in *2015 13th International Conference on Frontiers of Information Technology (FIT)* (IEEE, Piscataway, 2015), pp. 264–270

101. D. Jia, K. Lu, J. Wang, X. Zhang, X. Shen, A survey on platoon-based vehicular cyber-physical systems. IEEE Commun. Surv. Tutorials **18**(1), 263–284 (2016)

102. R. Hall, C. Chin, Vehicle sorting for platoon formation: impacts on highway entry and throughput. Transp. Res. C Emerg. Technol. **13**(5), 405–420 (2005)

103. P. Kavathekar, Y. Chen, Vehicle platooning: a brief survey and categorization, in *ASME 2011 International Design Engineering Technical Conferences and Computers and Information in Engineering Conference* (American Society of Mechanical Engineers, New York, 2011), pp. 829–845

104. K. Zhu, D. Niyato, P. Wang, E. Hossain, D. In Kim, Mobility and handoff management in vehicular networks: a survey. Wirel. Commun. Mob. Comput. **11**(4), 459–476 (2011)
105. G.I. Tsiropoulos, O.A. Dobre, M.H. Ahmed, K.E. Baddour, Radio resource allocation techniques for efficient spectrum access in cognitive radio networks. IEEE Commun. Surv. Tutorials **18**(1), 824–847 (2016)
106. E.Z. Tragos, S. Zeadally, A.G. Fragkiadakis, V.A. Siris, Spectrum assignment in cognitive radio networks: a comprehensive survey. IEEE Commun. Surv. Tutorials **15**(3), 1108–1135 (2013)
107. I.F. Akyildiz, W.-Y. Lee, M.C. Vuran, S. Mohanty, A survey on spectrum management in cognitive radio networks. IEEE Commun. Mag. **46**(4), 40–48 (2008)
108. D.W.K. Ng, E.S. Lo, R. Schober, Multiobjective resource allocation for secure communication in cognitive radio networks with wireless information and power transfer. IEEE Trans. Veh. Technol. **65**(5), 3166–3184 (2016)
109. L. Lyu, C. Chen, Y. Li, F. Lin, L. Liu, X. Guan, Cognitive radio enabled transmission for state estimation in industrial cyber-physical systems, in *2015 IEEE Global Communications Conference (GLOBECOM)* (IEEE, Piscataway, 2015), pp. 1–6
110. X. Cao, P. Cheng, J. Chen, S.S. Ge, Y. Cheng, Y. Sun, Cognitive radio based state estimation in cyber-physical systems. IEEE J. Sel. Areas Commun. **32**(3), 489–502 (2014)
111. Y.-C. Liang, Y. Zeng, E.C. Peh, A.T. Hoang, Sensing-throughput tradeoff for cognitive radio networks. IEEE Trans. Wirel. Commun. **7**(4), 1326–1337 (2008)
112. T.A. Khan, C.S. Hyder, A.A. Al Islam, Towards exploiting a synergy between cognitive and multi-radio networking, in *2015 IEEE 11th International Conference on Wireless and Mobile Computing, Networking and Communications (WiMob)* (IEEE, Piscataway, 2015), pp. 370–377
113. P. Mell, T. Grance, Draft NIST working definition of cloud computing-v15, 21. Aug 2009, vol. 2, pp. 123–135 (2009)
114. T. Dillon, C. Wu, E. Chang, Cloud computing: issues and challenges, in *2010 24th IEEE International Conference on Advanced Information Networking and Applications (AINA)* (IEEE, Piscataway, 2010), pp. 27–33
115. C. Zou, J. Wan, M. Chen, D. Li, Simulation modeling of cyber-physical systems exemplified by unmanned vehicles with WSNs navigation, in *Embedded and Multimedia Computing Technology and Service* (Springer, Dordrecht, 2012), pp. 269–275
116. X. Li, C. Qiao, X. Yu, A. Wagh, R. Sudhaakar, S. Addepalli, Toward effective service scheduling for human drivers in vehicular cyber-physical systems. IEEE Trans. Parallel Distrib. Syst. **23**(9), 1775–1789 (2012)
117. J. Wan, M. Chen, F. Xia, L. Di, K. Zhou, From machine-to-machine communications towards cyber-physical systems. Comput. Sci. Inf. Syst. **10**(3), 1105–1128 (2013)
118. J. Wan, D. Zhang, Y. Sun, K. Lin, C. Zou, H. Cai, VCMIA: a novel architecture for integrating vehicular cyber-physical systems and mobile cloud computing. Mobile Netw. Appl. **19**(2), 153–160 (2014)
119. J. Wan, D. Zhang, S. Zhao, L. Yang, J. Lloret, Context-aware vehicular cyber-physical systems with cloud support: architecture, challenges, and solutions. IEEE Commun. Mag. **52**(8), 106–113 (2014)
120. A. Rakotonirainy, Design of context-aware systems for vehicle using complex systems paradigms (2005)
121. S. Al-Sultan, A.H. Al-Bayatti, H. Zedan, Context-aware driver behavior detection system in intelligent transportation systems. IEEE Trans. Veh. Technol. **62**(9), 4264–4275 (2013)
122. J. Santa, A.F. Gomez-Skarmeta, Sharing context-aware road and safety information. IEEE Pervasive Comput. **8**(3), 58–65 (2009)
123. K.K. Venkatasubramanian, S. Nabar, S.K. Gupta, R. Poovendran, Cyber physical security solutions for pervasive health monitoring systems, in *User-Driven Healthcare: Concepts, Methodologies, Tools, and Applications* (IGI Global, Hershey, 2013), pp. 447–465
124. M. Bouet, A.L. Dos Santos, RFID tags: positioning principles and localization techniques, in *Wireless Days, 2008. WD'08. 1st IFIP* (IEEE, Piscataway, 2008), pp. 1–5

125. Y. Khan, A.E. Ostfeld, C.M. Lochner, A. Pierre, A.C. Arias, Monitoring of vital signs with flexible and wearable medical devices. Adv. Mater. **28**, 4373–4395 (2016)
126. T. Shah, A. Yavari, S.S. Karan Mitra, P.P. Jayaraman, F. Rabhi, R. Ranjan, Remote health care cyber-physical system: quality of service (QoS) challenges and opportunities. IET Cyber-Phys. Syst. Theory Appl. **1**(1), 40–48 (2016)
127. M. Chen, Y. Ma, Y. Li, D. Wu, Y. Zhang, C.-H. Youn, Wearable 2.0: enabling human-cloud integration in next generation healthcare systems. IEEE Commun. Mag. **55**(1), 54–61 (2017)
128. M.S. Hossain, M.A. Rahman, G. Muhammad, Cyber–physical cloud-oriented multi-sensory smart home framework for elderly people: an energy efficiency perspective. J. Parallel Distrib. Comput. **103**, 11–21 (2017)

Chapter 4
A Framework for Speculative Job Scheduling on Mobile Cloud Resources

Ansuman Banerjee, Himadri Sekhar Paul, Arijit Mukherjee, Swarnava Dey, and Pubali Datta

4.1 Introduction

Recent advances in mobile technology have enabled immense penetration of these devices in the common market. At the same time, ancillary businesses revolving around mobile devices have attained a boost of similar magnitude. With all these developments both in the technical front and in business, mobile devices are poised to revolutionize the personal computing landscape. The mobile application market alone is estimated to reach US$77 billion by the end of 2017 [1]. Although computing capacity of mobile devices has significantly increased over the past decade [2], so did the computation demand from their users. The complexity and therefore the computation requirement of the mobile applications have increased in similar pace over time to keep up with user expectations. A typical smartphone performs various tasks, including management and responses to user interactions. Typically there are several computation hungry tasks which run in the background to enhance user experience with its device. Since a mobile device is a very personal device, user's experience with its device is of prime importance.

This work was done during Pubali's association with TRDDC, Pune, India

A. Banerjee
ACMU, Indian Statistical Institute, Kolkata, India
e-mail: ansuman@isical.ac.in

H. S. Paul (✉) · A. Mukherjee · S. Dey
TCS Research and Innovation, Kolkata, India
e-mail: himadriSekhar.Paul@tcs.com; mukherjee.Arijit@tcs.com; swarnava.dey@tcs.com

P. Datta
Department of Computer Science, University of Illinois Urbana Champaign, Champaign, IL, USA
e-mail: pdatta2@illinois.edu

© Springer Nature Switzerland AG 2020
S. Hu, B. Yu (eds.), *Big Data Analytics for Cyber-Physical Systems*,
https://doi.org/10.1007/978-3-030-43494-6_4

103

Choice of applications running on mobile devices is based on user preferences. The pattern of usage of a device is also specific to a user. Researchers have attempted to discover user specific patterns in usage of a device. Study of such patterns is important in various contexts, including network usage [3], battery charge decay characterization [4], etc. Shye et al. introduced a Markov decision process-based model to capture user activity [5]. In this paper, we model the usage pattern of a device as a state transition system. Our model is simple, yet effective, in the contexts for which it was used in our experiments. We used this model in scheduling background tasks in a mobile device and also in the context of collaborative computing in mobile cloud.

Similar to a regular desktop computing environment, a smartphone also performs several routine jobs in the background to keep its computing environment up-to-date and enhance user experience with its device. A smartphone performs system updates in regular intervals, runs virus scans at regular intervals, builds file indexes, mines logs (like call logs) to build knowledge bases, etc. There may be several such background activities, which are important and yet low-priority jobs for the scheduler, which help in creating a comfortable computing environment for its user. Most of these background activities are resource hungry and can consume considerable amount of CPU cycles while running. It is not an uncommon experience that such jobs trigger off in uncanny hours when the user is very active with its device, resulting in delayed response from the application the user is using. There are usually trapdoors available for a user to specify a schedule for these activities individually. Such schedules are static and do not cater to the dynamic nature of usage of the device. In this paper, we propose to employ the usage model to intelligently suggest when to run a background task, such that demand for resources is evenly distributed temporally, resulting in better user experience with the device.

In this paper, a state transition model of user usage pattern is used as a guidance for estimating execution time of a given task. The model is also used to determine a schedule of the task in question, such that sufficient resource is available to the background task without overloading the system and also to ensure that the user experience with the device is not affected. This basic technique can be applied in a different form, in the context of collaborative computing involving mobile devices. With global penetration of mobile devices in the commercial market, the count and capacity of the devices are in the rise. To harness the free computation cycles of these devices, several collaborative computing frameworks have been developed, namely Hyrax [6], Misco [7], Serendipity [8], etc. In this paper, we adopt a localized mobile grid setting where the devices are accessible through a WiFi connection. We examine the problem of computation scheduling and workload management for improving timing/energy performance. We consider a private company infrastructure with a gateway device and a mobile grid, where the gateway device is expected to host and assimilate an information database on which some computation need to be executed. The gateway device needs to decide on a schedule of computation and a selection mechanism so as to engage the mobile devices and utilize their donated computation cycles. The primary objective of driving this selection is to be able to finish execution of the application at the earliest possible time.

The paper is organized as follows. In Sect. 4.2, we present a model of pattern of usage of a mobile device and also present a discussion on how to extract such a model from real usage traces. Section 4.3 presents a scheduler, augmented by the usage model, such that the usage experience with the device is not affected by execution of background jobs and present some experiment results in Sect. 4.4. We use the usage model in a mobile and cloud collaborative computing setting and present motivation for the same in Sect. 4.5. We present such a collaborative model based on bidding in Sect. 4.6 and our experiments in Sect. 4.7. Finally, Sect. 4.9 concludes this paper.

4.2 Modelling Usage Patterns

A mobile device is probably the most personal device that its owner carries and the device is most personalized by its owner. The operation of such a device carries its owner's signature. It has been observed that there is a pattern of use for each device owner and based on this assumption, researchers have tried to model the usage pattern of a device on various aspects. One of the objectives of such a model is to predict the temporal variation of certain aspects of the device. In this paper, our objective is to estimate the execution time of a given task utilizing free computation cycles of the device. We model the mobile device as a *probabilistic finite state machine with average permanence* (PFSM-AP). The PFSM-AP model is defined as a tuple $\mathscr{U} = < \mathbf{S}, \mathscr{I}, \mathbf{T}, \lambda, \mathbf{H}, \mathbf{C} >$ where,

- $\mathbf{S}$ denotes the set of states.
- $\mathscr{I}$ is the set of external events.
- $\mathbf{T}$ denotes the transition function $\mathbf{T} \subseteq \mathbf{S} \times \mathbf{S} \times \mathscr{I}$.
- λ is the transition probability function, defined as

$$\lambda \left(s_i, s_j \right) = p_{ij} \quad \text{where,} \ s_i, s_j \in \mathbf{T}$$

such that, for each s_i the sum of the transition probabilities on its outgoing edges is 1.
- $\mathbf{H} : \mathbf{S} \to \mathfrak{R}$ is the average permanence function, defined as,

$$\mathbf{H} \left(s_i \right) = t_i \quad \text{where,} \ s_i \in \mathbf{S}, t_i \in \mathfrak{R}$$

$\mathfrak{R}$ is the set of reals.

- **C** : **S** $\rightarrow$ [0, 1] is the CPU availability fraction or equivalently availability percentage.

The objective of the model depicted as above is to characterize a device based on its free CPU cycles. However, the model is not restricted to this feature only and can easily be extended to include any other feature of interest. We restrict our model to free CPU cycles since we use only this feature in our estimation of execution time, as described below.

4.2.1 Battery Decay

The source of energy of a mobile device is of primary consideration and the rate of decay of charge plays an important role in its schedule. For example, if an application takes such a long time to execute in the mobile that it is likely to exhaust all its energy, it is not advisable to schedule that task. The primary source for battery power consumption in a mobile phone has been identified as its screen and communication modules [9, 10]. There have been several interesting proposals of optimizing power consumption in mobile devices based on energy profiles of the communication devices and scheduling data volume transfers [11–13]. Authors in [14] analyze and present power consumption results based on usage pattern of mobile devices. The authors claim that primary sources of the power decay in mobile devices are active screen and CPU. Their study reveals that majority of the usage patterns have long intervals between two subsequent active screens. Power modelling is an involved field of study involving several sub-areas like architecture, operating system, software engineering, etc. Instruction level power models have been proposed in literature to estimate the power consumption by applications [15, 16]. The objective of these models is to accurately characterize an application for their power consumption which can be used for power optimization of the application. Authors in [4] propose a method for usage pattern-based estimation of battery power. They use an auto-regression model on logged usage patterns to predict device power usage. However, in our context, we only require a coarse-level estimation of power consumption by an application, without performing such expensive profiling techniques. In this work, our objective is to utilize free computation cycles of a mobile device and we assume that computation power of CPU does not change due to dynamic voltage and frequency scaling (DVFS) level. In this work, we use a simple model to characterize decay of battery charge in mobile devices. For the sake of simplicity of illustration, we assume here that the expenditure in battery energy by an application is linear in the number of CPU cycles it consumes.

$$\mathbf{B} = \beta \times n$$

where n is the number of CPU cycles required by an application, **B** is the energy required for n cycles, and β is a known constant

4.2.2   Generation of PFSM-AP Model

Generation of the usage model was the first phase of our experiment. As part of model generation, we chose a relatively small history of usage of a device to model recent usage patterns of the person. Usage is known to vary widely over time, since interests and need of the person change over time. During this phase, we captured usage patterns of seven mobile devices owned by seven of our employees who volunteered to donate their devices for our experiments. The description of the devices is shown in Table 4.1. These devices include two Sony Xperia devices, three Samsung Galaxy devices, and one each of Google Nexus and Micromax Canvas devices. During this phase of the experiment, we worked towards building the PFSM-AP models of the mobile devices participating in our experiments. We developed and installed a small Android application which can collect device usage trace data (like free memory and CPU usage) every 5 s and log into the devices. The users carried this application, active in their devices, and the application gathered data for approximately a month. We then collected this trace data and analyzed them offline to build their corresponding PFSM-AP models. The trace data for the last day, however, was not considered for building the model, but was replayed during our experiments with this model. We extracted the percentage of free CPU cycles only from the data and applied clustering to build the PFSM-AP. The percentage of free CPU cycles of a state was calculated as the mean of the data points. Once the state transition model was built, we also derived, from the log, the durations the device remained in a certain state. The permanence value of the state was computed as the average of these durations. The method is outlined as Algorithm 1.

The clustering heuristic is a form of density-based clustering and works itera-tively to refine a set of initial clusters. The heuristic has two phases. The first phase of the clustering heuristic creates an initial set of clusters based on the feature of the CPU availability value (data gathered as the percentage of free CPU cycles). In the initial clustering phase (line 1 of Algorithm 1), we create some initial clusters based on the range of CPU availability value. In this phase, we create a set of *set of clusters*, based on different bucket sizes. A bucket is defined on a range of values for a feature, in this case, the CPU availability value. For example, consider buckets of size 2. Data points with CPU availability value in the range of [0, 1] (i.e., the first bucket) are put into one cluster, points in the range of [2, 3] (i.e., second bucket) are put into a different cluster, and so on. Since the CPU availability values are in

Table 4.1 Configuration of mobile devices used in our experiment

Device name	Model	OS Android Ver	CPU Core @ clock speed	Memory
Samsung Galaxy	GT-S6802	2.3.6	Single Core @ 832 MHz	512 MB
Sony Xperia L	C2104	4.1.2	Dual Core @ 1 GHz	1 GB
Micromax Canvas 2+	A110Q	4.2 (Jelly Bean)	Quad Core @ 1.2 GHz	1 GB
Google Nexus	Nexus-4	4.2 (Jelly Bean)	Quad Core @ 1.5 GHz	2 GB

Algorithm 1: PFSM-AP model determination

---

begin

    // Phase 1: Initial Clustering - Empirical Analysis

1   **for** $i \leftarrow a$ **to** b **do**

        cluster data points in buckets of size i;

        $\delta_i \leftarrow$ deviation of cluster size values;

2   Choose i^* *s.t.* δ_{i^*} is the highest in $\{\delta_i : a \leq i \leq b\}$;

3    $\mathscr{C} \leftarrow$ cluster data points in clusters of size i^*;

4    $\{CC_k\} \leftarrow$ Compute cluster centers of $\mathscr{C}$;

5

    // Phase 2: Refinement and Reclustering

6   **while** *No change in cluster composition* **do**

7   Recluster data points around cluster centers $\{CC_k\}$ based on the distance of a point from cluster centers;

8   Remove a cluster if the size of the cluster is less than $\dfrac{\text{No of data points}}{|\mathscr{C}|} \times \frac{\sigma}{100}$;

9   Reassign data points of removed clusters to existing clusters based on the distance of a point from cluster centers;

10        $\{CC_k\} \leftarrow$ Compute cluster centers of $\mathscr{C}$;

11   Each cluster is a state and the mean value is the percentage of the free CPU cycles;

12

    // Phase 3: Computation of Transitions and Transition Probabilities and Creation of PFSM-AP Model

13   Traverse the data and compute average time in a state;

14   Traverse the data and compute number of transitions for all pairs of states;

15   Compute the transition probability of an edge $s \rightarrow d$ as the fraction of transitions from state s to d against all transitions out of state s, i.e. $\dfrac{\text{No. of transitions from } s \text{ to } d}{\sum_{\forall p} \text{No. of transitions from } s \text{ to } p}$

---

the range [0, 100], we can construct 50 buckets and therefore 50 initial clusters can be constructed for a bucket size of 2. Then we compute the deviation of the CPU availability value for each cluster. Such sets of cluster-sets are created for all bucket sizes in the range of $[a, b]$. The exact values for a and b are chosen depending on the number of observations. It is intuitively obvious that a very high bucket size will create too few clusters. Out of these sets of cluster-sets, we choose the one where the deviation is the highest (line 2 of Algorithm 1). Empirically the chosen cluster-set captures densely packed points into a single clusters. Such a set serves as an initial set of clusters to be refined in the subsequent phases of the heuristic.

In the next phase, we refine the clusters and their memberships. The refinement is done by modification of clusters with low membership count. Any cluster having number of data points less than a threshold is removed. The threshold is defined as σ percent of average membership count of all the clusters. The value of σ was taken as a parameter to the algorithm. Typically the value of σ is small and in our experiment σ was chosen as 5 which indicates that clusters with less than 5% of the average cluster size are modified. Then a cluster refinement is done as follows. All the data points are reclustered based on their distances from the centers of the surviving clusters. Once reclustering is complete, the cluster centers are recomputed.

The last phase of the algorithm is the construction of the PFSM-AP model. The process is straightforward. Each of the clusters constructed in phase-2 represents a state in the model. The clusters are enumerated and the enumeration values are the cluster-ids or state numbers. Each of the entries of the mobile trace data now can be annotated with a state number. Since the trace data entries also contain time-stamps, it is easy to determine the duration the device stays in a certain state. For this analysis, the computation of average permanence for each state is also simple. Once the trace data is annotated with state numbers, it is also easy to define state transitions and assign a transition probability, as shown in line 15. Figure 4.1 shows a part of the CPU usage pattern of one of the users, and the PFSM-AP model constructed thereafter is shown in Fig. 4.2.

Fig. 4.1 CPU usage pattern

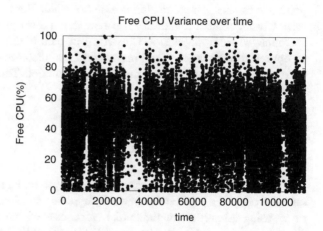

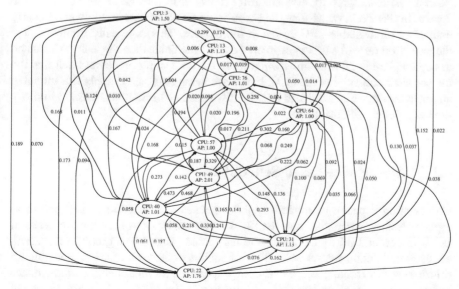

Fig. 4.2 PFSM-AP model

4.2.2.1 Analysis of the Heuristic

The heuristic presented as Algorithm 1 has two loops. The for loop at line 1 indexed by variable i is bounded by variables a and b. We now show that the while loop at line 6 executes finitely many times. The while loop is executed until no data point is reassigned to a different cluster (through cluster modification). Assignment of a data point to a cluster changes due to two effects: (1) One or more clusters are modified (line 8), and (2) the cluster center shifts, changing the membership of the data points (line 9). Let us consider the first case. For any given $i^* : a \leq i^* \leq b$, the number of initial clusters is bounded. Since the algorithm only allows small clusters to be destroyed and no new cluster is created inside the while loop, the cluster removal step (line 8) is bounded. Therefore, the number of cluster reassignments for a data point due to cluster modification is also bounded. Now, let us consider the second case. Cluster centers can shift when new data points join a cluster or data points are removed from the cluster. Clustering of data points around cluster centers is an optimization problem where minimization of distances of data points from their corresponding cluster centers is the objective function. This also executes for finitely many iterations.

4.2.3 Execution Estimation Model

We assume that the execution time of a given job on a given device architecture is known a priori (i.e., can be computed by dynamic simulation against a given dataset or by using established methods like worst case execution time estimation [17]). Such an estimate typically assumes full (100%) utilization of the resources in the device. In this work, we assume the execution time of the task is solely and linearly dependent on available CPU cycles in the devices. This essentially implies that if the execution time of a task is estimated as 10 time units, then the task is estimated to be complete in 20 time units when there are competing processes such that the job can avail only 50% of the CPU. In the following section, we apply this model to augment the device scheduler to intelligently schedule background tasks to improve user experience with the device.

4.3 Usage Model-Based Scheduler

Computation capacity of modern mobile devices is increasing and is poised to replace desktop computing devices [2]. Mobile devices run broadly two types of tasks, foreground jobs with which the user interacts and background jobs which are essentially maintenance tasks. The schedulers of these OSes are responsible for prioritizing foreground jobs so that users experience minimum delay. Maintenance tasks, like virus scans, system updates, building file index, etc. which are not part of typical usage of devices, run as low-priority jobs in the background, consume much less system resources, yet are important for the operation of the device. The usage model of a device, described in Sect. 4.2, captures its resource availability under regular usage by its owner and its variation over time. Many of the background tasks are periodic tasks and usually the OS provides means to the user to schedule them, such that they are triggered when the device is relatively free. Yet it is a common experience that such tasks are triggered at times when the user is in active use with the device. This results in a race for resource acquisition and results in slower response. To improve user experience in such cases, it is important to determine a schedule of such tasks when the usage of resources is likely to be low. A scheduler can leverage the usage model to intelligently schedule background tasks such that the user can more comfortably use its device without being affected/annoyed. In this section, we present a model of a scheduler which is aware of the usage pattern of the device.

4.3.1 Model of Scheduler

We consider a mobile device with usage pattern $\mathscr{U} =< S, \mathscr{I}, T, \lambda, H, C >$ as presented in Sect. 4.2. There is a set of background jobs to be scheduled by the scheduler, opportunistically. Each of these jobs have deadlines associated with them, and we want to maximize the number of jobs which successfully complete within their deadlines. Let $J = \{J_1, J_2, \ldots J_n\}$ denote a set of background jobs. The deadline of each of the jobs is denoted as $\mathscr{D}(J_i)$. Also we denote by $\mathscr{E}(J_i)$ the estimated execution time of the task on the device.

The selection process of a background job from the set J for execution is based on the present computation state and battery state of the device. The objective of the process is to identify a job which will fit the device's energy budget and deadline of the job, without hindering the operations done by the user on the device. We present an outline of the steps for the selection process below.

1. Sort jobs in J in increasing order of their deadlines.
2. Choose a path, $P =< s_0, s_1, s_2, \ldots s_m >$ in PFSM-AP model, such that for every edge (s_{i-1}, s_i) in P, $\lambda(s_{i-1}, s_i)$ is highest among all the outgoing edges from s_{i-1}.
3. Selection of job: for all jobs P_i, taken sequentially from the sorted job list, do the following:

   - Projection of execution: Calculate the time required to execute J_i on path P such that,

     (a) only α-fraction of free CPU of every state in P is used for execution of J_i
     (b) execution of J_i on path P completes within $\mathscr{D}(J_i)$
     (c) decay of battery on this execution time does not result in complete drainage of the battery.

   Schedule J_i immediately if all the above conditions are met.

We preserve α-fraction of the free resources as buffer in each state of the PFSM-AP and this parameter can be used to control free resource usage by background jobs. Therefore, a job can only utilize α-fraction of the free computation cycles of the device so that the user's experience with its device is not degraded. Since a device cannot offer a steady computation power to a background job, since it ensures that user's jobs have higher access priority over CPU, the background job executed in a device experiences variable computation power. The classical version of this problem is equivalent to *open shop scheduling* with multiple jobs and single machine, and it is known to be NP-hard [18]. In contrast to the classical open shop scheduling problem, the computation power of the machine varies over time.

The PFSM-AP model presented here models the varying computation power available for a background job. Each state of the PFSM-AP model represents available resource in that state. Given an initial state in the PFSM-AP model which represents the present computation state of the device, the selection process needs to refine $\mathscr{E}(J_i)$, the estimated run-time of the job J_i on the device with no other computation load. We present an on-line heuristic as Algorithm 2, which examines the jobs, sorted in their ascending order of their deadlines (i.e., job with the nearest deadline as higher priority for scheduling) and estimates whether the job can be completed within its deadline before the battery is drained out. The estimation process is based on the estimation of execution time of J_i, which shares CPU with other jobs being initiated by the user all of which are treated as higher priority jobs than J_i.

For the process of refinement of execution time of a given job, one has to determine a path in the model, given an initial state. Refinement, in our case, is essentially a projection of $\mathscr{E}(J_i)$ on the path, based on the available computation cycles in the path and is computed in lines 4–6 of Algorithm 2. Therefore, the choice of the path determines the accuracy of the refined estimation of the execution time. The edges of the PFSM-AP model are annotated with probability of state transition by the CPU. For each of these paths, we are interested in the shortest path-prefix which can accommodate the computation requirement of the job. Since we are only interested in a path-prefix which the device is most likely to follow, we need to choose one which has the highest probability value, computed as the product of the constituent edges. Evidently the problem is computationally hard but we require to solve it efficiently to be able to quickly choose a job from the list. The heuristic presented here is based on the assumption that user behavior on its device is frequently repeated. So we always follow the outgoing edge having the highest probability value (line 7).

4.3.1.1 Analysis of the Heuristic

The heuristic has two loops nested. The outer loop at line 1 takes one job from a list and refines its execution time in the inner loop at line 3. The execution for the

---

Algorithm 2: $\alpha Sched_1$: Scheduling of a job with one deadline

---

    input : $\mathbf{J} = \{J_1, J_2, \ldots J_n\}$: Set of jobs to be executed
    input : $\mathscr{U}$: PFSM-AP model of the device
    input : $\mathbf{s}$: Present state of the device
    input : $\mathscr{B}$: Estimated battery charge level
    output: X : Job to be scheduled
    begin

	$\mathbf{J'} \leftarrow$ Sort $\mathbf{J}$ in increasing order of their deadline i.e., $\mathscr{D}(J_i)$;
1	**foreach** $J_i \in \mathbf{J'}$ **do**
2	$T \leftarrow \mathscr{E}(J_i)$: estimated computation time of J_i;
	$tm \leftarrow 0$ // Wall clock : Initialization
3	**repeat**
4	$c \leftarrow (1 - \alpha) \times \mathbf{C(s)} \times \mathbf{H(s)}$;
5	$T \leftarrow T - c$;
6	$tm \leftarrow tm + \mathbf{H(s)}$;
7	$e \leftarrow (\mathbf{s}, r)$ where $\lambda(\mathbf{s}, r)$ is maximum among all outgoing edges from $\mathbf{s}$;
	$\mathbf{s} \leftarrow r$;
	// Exceeds deadline - not feasible
	if $tm > \mathscr{D}(J_i)$ **then** continue at 1
	// Exhausts battery - not feasible
	if $\mathscr{B} \le (\beta \times tm)$ **then** continue at 1
	until $T \le 0$;
8	Schedule J_i for execution and return;

---

outer loop is bounded by $|\mathbf{J}|$. The inner loop essentially traverses the PFSM-AP model from the initial node, given as input to the heuristic, following the edge with the highest probability value. The loop is terminated when the remaining execution time of the job, T, drops beyond 0. In each iteration, the value T monotonically decreases. The PFSM-AP model can possibly include at-most one state with 0% available computation capability, due to clustering process. Also by construction of the PFSM-AP model, no state can have a self-loop. A self-loop indicates that the device remains in the same state after a duration. However, the whole duration for which the device remains in a state is computed as permanence for that duration. By model construction, the average of all permanence values is computed and stored as *average permanence* for the state. Given a self-loop free graph of the PFSM-AP model, at-most one state (with 0% CPU availability) at which T value decreases is chosen and the value of T monotonically decreases in each iteration of the inner loop. Thus, the algorithm terminates in a finite number of steps. In the next section, we present comparison of our schedule with shortest job first (SJF) schedule [19] with priority in a simulated environment.

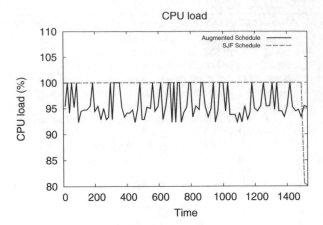

Fig. 4.3 Comparison of usage pattern augmented scheduler

4.4 Result: Usage Model-Based Scheduling

We carried out simulation experiment to compare our heuristic with the SJF schedule. Our technique utilizes the usage pattern captured during our first phase of experimentation described in Sect. 4.2.2. We used the PFSM-AP model to augment our scheduler. We simulated the mobile devices and implemented our heuristic as an augmentation of the shortest job first scheduler. The background jobs were simulated to be of varying duration from the range [10, 50] s with their deadlines to be twice as their execution time, and we simulated 100 such jobs of low-priority. We used $\alpha = 0.2$ as specified in Algorithm 2.

Figure 4.3 shows a comparison of our heuristic with the SJF scheduler. It is evident that a scheduler, which statically schedules all background jobs, always keeps the device busy. As a result, the user may experience delayed responses from the device. Our augmented scheduler conservatively uses the free computation cycles of the device such that foreground jobs have some buffer cycles left and therefore the user can use the device more comfortably. We can see from the figure that CPU utilization is not always 100% and has some room to cater to additional resources demanded by the foreground jobs and as a result the user is expected to experience a better response from the device. As expected, in the case when our scheduler is used, the background job takes longer time to complete.

4.5 Usage Pattern for Mobile Grid

Cloud computing involving mobile devices is an active area of research which deals with utilization of mobile devices in collaborative computing with cloud's back-end computing infrastructure. This paradigm of computing has two facets: in one, mobile devices are used as computing resources by the back-end infrastructure and, in the other, mobile devices use the back-end infrastructure to augment their computing capacity. Motivation for the first approach is the growing capacity and market for smart devices, which is stimulating the prospect of utilizing them as computing resources [20, 21]. Recent studies on the computing capacity of mobile devices claim that their computing power is comparable to that of desktops [2]. Frameworks like Hyrax [6], Serendipity [8] attempt to exploit the free computing capacity of mobile devices. The later one is a more traditional approach to mobile cloud computing (MCC) where devices utilize the computing power of back-end infrastructures by offloading some of its tasks to the back-end. This has led to several proposals of MCC for collaborative execution for executing compute-intensive work-flows [22–25]. The MCC paradigm has attracted considerable attention in both academia and the industry community in recent times. In this paper, we present a model and system for utilization of free computation cycles of mobile devices in MCC.

Several challenges remain to engage a mobile device as part of a computing infrastructure [26]. Some of these challenges are limited communication bandwidth, energy constraints, memory capacity, intermittent availability of resources like network or CPU, proper incentive schemes against utilization, security, privacy, etc. In a controlled environment, some of these constraints can be addressed adequately in order to utilize the computation capacity of the mobile devices. For example, many of the reputed commercial organizations distribute smartphones among their senior employees [27]. In such a corporate environment, it is possible to make it a policy that such phones be used for computation for the benefit of the company's infrastructure. Such a device can be used by the infrastructure whenever the device is present in the premises of the organization and is connected to the internal communication network. In such a scenario, the issues of communication reliability and cost, security, and privacy are mitigated. To encourage such an environment, the organization may as well provide incentives in suitable forms. In the company of the authors, reward points are awarded for additional participation in company tasks (apart from regular assigned duties) and these points can be redeemed against purchases promoted by the organization.

In this part of the paper, we adopt a simple localized mobile grid setting where the devices are accessible through a WiFi connection, and examine the problem of computation scheduling and workload management for improved timing/energy performance. We consider a private company infrastructure with a gateway device and a mobile grid, with the gateway device hosting and assimilating an information database on which some computation need to be executed. The gateway device needs to decide on a schedule of computation and a selection mechanism so as to engage the mobile devices and utilize their donated computation cycles. The primary objective of driving this selection is to be able to finish execution of the application at the earliest possible time.

The gateway device is enabled with a task off-loader which is the controller of the task selection framework. When the off-loader wants to execute a task (in the form of a downloadable application), it invites bids from the owners of all devices connected to the off-loader. Additionally, the off-loader announces a deadline by which the computation has to finish. Associated with the task is a suitable reward to be earned by the winning bidder and also a penalty if the winner fails to deliver the task in time. Each owner, intending to participate in the bid, executes a pre-installed analysis agent on his device. The agent takes as input the advertised task and the deadline associated with the task, and comes back to the owner with an advice whether to bid or not, on the basis of its estimation of the execution time. In this paper, we consider the estimation of execution time of a task with various levels of information available about the device usage pattern. In our architecture, the mobile devices are active agents, who learn and build models of their owner usage patterns. The owner places a bid only if the estimated execution time is less than the advertised deadline of the job. The bid is the promised completion time within which the corresponding device can complete the advertised task. The off-loader can possibly select one of the bidding agents for offloading the task based on some criterion. In the simplest case, it may choose the one with earliest promised completion time and offload the task to the selected device. We assume that the owners are rational (aware of penalty) and honest (no false bids). The interesting activity from the device's perspective is to analyze how/when/what to bid for, while designing the selection and scheduling mechanism is the off-loader's challenge.

4.5.1 Motivation for This Work

In this section, we present an example to illustrate the need for modelling a mobile device for its usage. A mobile device has various operational modes in its usage cycle. For example, when a user attends to a call, its communication modules are busy, when he listens to music or radio, its audio system is busy, and when he watches a movie, its GPU remains busy. Manufacturers of mobile devices usually specify an operating model of their devices. An operating model is a state transition system where the states represent some high level operation modes (e.g., charging, audio on, network on, etc.) with average/maximum/minimum resource usage estimates when the device operates in that particular state, and possible interstate transitions. The operating condition of the device in these states can be attributed to its usage of processor, memory, cache, priority of the running jobs, battery power state, etc. The transitions in such a system are triggered by user interaction of the device and usage of device resources by various applications running in the system. In this paper, we extend this model to a usage-induced operating state model, a transition system based on the operating model and, additionally, specialized by the usage pattern of the device owner. In the context of exploiting a mobile device in our setting, we are interested in the availability of different resources in the device to utilize it for running an external computation. For simplicity, we assume here the states in the usage model of a device are characterized only by the percentage of CPU available.

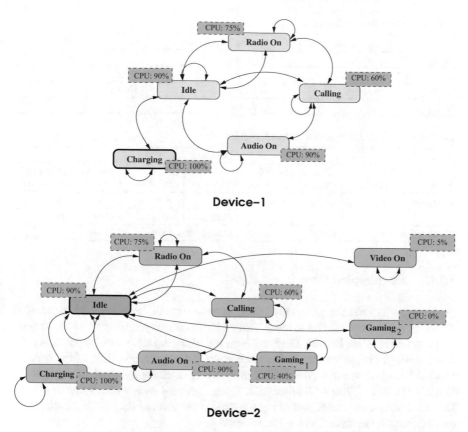

Fig. 4.4 Usage model with CPU availability

As an example, we consider here a simple case of two mobile devices and one task to be offloaded to one of the devices. The task has a deadline of 60 time units. Each mobile device needs to estimate its bid based on its operational state model, as depicted in Fig. 4.4. The events triggering the transitions are not shown in the figure, since they are not required for presentation of these examples. Each state in the state model is annotated with the fraction of CPU available for external computation at that state, which can be used for executing the external task. For the sake of simplicity, we assume here that the device takes one of the out-bound transitions from its current state, including the self-loop, after every unit time. In other words, the device stays at each state for one unit of time, executes one of the outgoing transitions from the present state, and moves to the next state (may be same as the current one) where it stays for one more unit, and this continues. We assume such transitions are instantaneous (Table 4.2).

Table 4.2 Execution on Device-1

State	CPU availability	Time in the state	Effective execution
charging	100%	1 sec	1 sec
idle	90%	10 sec	9 sec
calling	60%	50 sec	30 sec
Completion time:		61 sec	

Table 4.3 Estimated completion time with best transition

Device-1				Device-2			
State	CPU availability	Time in the state	Effective execution	State	CPU availability	Time in the state	Effective execution
charging	100%	40 sec	40 sec	*idle*	90%	10 sec	9 sec
				charging	100%	31 sec	31 sec
Completion time:		40 sec		Completion time:		41 sec	

4.5.1.1   The Simplest Case

Both the devices have an estimate of the execution time of the advertised application on their architecture. Let us assume both of them come up with a value of 40 time units. When bids are invited, Device-1 is in the *charging* state and Device-2 is *idle*. If the devices always remain in the same state, the completion time of the task on Device-1 is 40 time units (100% CPU availability in *charging* state), while that for Device-2 is $40 \times \frac{100}{90} = 44.44$ time unit. Thus, Device-1 bids with a value of 40 and Device-2 bids with 44.44. Assuming the *off-loader* awards the job to the one with earlier completion time, Device-1 is selected.

4.5.1.2   A More Realistic Scenario

In a more realistic setting, each device is expected to transition away from its current state during the job execution and therefore cannot guarantee constant CPU availability. In such a setting, the device can explore all possible paths in its state graph and optimistically choose a path which provides the best estimated completion time. Such a path obviously would go through states with high CPU availabilities. For example, Device-1 would consider the path involving only the *charging* state, which always guarantees it 100% CPU availability for the external job and can bid with value 40. On the other hand, Device-2 would consider the path from *idle* to the highest CPU available state, i.e., *charging*. Table 4.3 shows the estimated completion times in this case and the *off-loader* may again select Device-1 for offloading.

Typically a state transition is triggered by external events, for example, incoming call, user's operation, etc. The execution paths chosen for bid as depicted above are therefore too optimistic. Consider the following scenario. At the time of execution of the external task, Device-1 remains in *charging state* for 1 time unit, in *idle* state for

10 time units, and then moves to the *calling state* and remains there. The execution completion time is shown in Table 4.2. The device thus completes 40 time units of computation in an effective duration of 61 time units and exceeds the deadline. This shows that only the best timing is not always a good candidate to decide on the bid, since a penalty is involved. A rational owner should ideally take this into account. On the other hand, a pessimistic strategy considering a maximal timing path may yield a completion time beyond the deadline. In either of the strategies, the path chosen for computation of a bid may not be the actual path taken during execution of the external task.

A more realistic estimate can be obtained by considering paths induced by the average usage by the user. To incorporate this, we further associate with each state an *average permanence* (AP) value [28] and a transition probability on each outgoing edge. AP implies the average time the device stays in the associated state. The revised model of the devices is depicted in Fig. 4.5. Now we apply the same

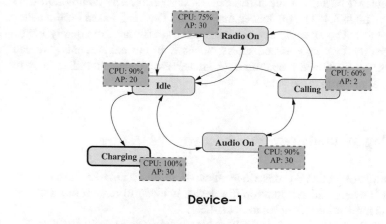

Device–1

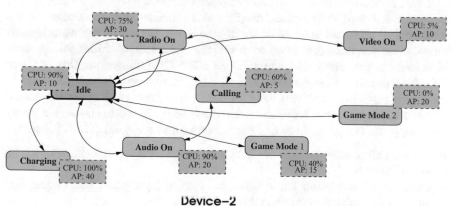

Device–2

Fig. 4.5 Usage model with average permanence

Table 4.4 Estimated completion time with AP

Device-1				Device-2			
State	CPU availability	Time in the state	Effective execution	State	CPU availability	Time in the state	Effective execution
charging	100%	30 s	30 s	*idle*	90%	10 s	9 s
idle	90%	10.1 s	10 s	*charging*	100%	31 s	31 s
Completion time:		41.1 s		Completion time:		41 s	

optimistic bid selection method based on the AP on states, assuming all transitions are equally likely. Also for each path we compute the probability of taking the path. This probability is a measure of confidence of the device taking that path. Device-1 chooses the path *charging* → *idle* which is associated with confidence value of 1 (since there is a single transition from *charging* state). The best confidence value ($1/8 = 0.125$ considering each of the 8 outgoing transitions are equally likely) for Device-2 occurs for the path *idle* → *charging*. The completion time is computed in Table 4.4. The off-loader may choose Device-2 based on the better bid proposed by it. The example above assumes the transitions are equally likely. However in reality, they may not be so, as we show in our experiments. We can learn the transition likelihood probabilities from user usage data and utilize them to enrich the bid above with these values.

4.6 Bidding Methodology

The objective of the PFSM-AP described above is to characterize a device based on its free CPU cycles and the duration the device is likely to remain in a state, as depicted in the motivating examples in Sect. 4.5.1.

Given a mobile device D_i with a PFSM-AP model $\mathcal{U}_i =< \mathbf{S}_i, \mathcal{I}_i, \mathbf{T}_i, \lambda_i, \mathbf{H}_i, \mathbf{C}_i >$, a task J with its dataset and deadline Δ, the device needs to calculate its bid which can be presented to the off-loader by the owner. Let us assume the task needs an estimated execution time of w_i on this device. As discussed earlier, this estimate is agnostic to the state model and assumes 100% CPU utilization. This is where PFSM-AP provides a better estimate. If $w_i > \Delta$, there is no point for the device to participate in the bid (intuitively there is no path in which the task can be completed within deadline even with 100% utilization all-through). The case is interesting only when $w_i \leq \Delta$. The principle behind our bid computation algorithm is as follows:

- Examine all possible paths in the PFSM-AP graph from the state the device is in, at the time when bids are invited.
- Compute expected completion times on each of these paths considering that states in the path have different CPU availability.
- Exclude paths where the expected completion time is greater than the deadline.

- Exclude paths where the corresponding confidence value is less than some pre-determined threshold.
- Determine a path which meets the deadline best and present the expected completion time on that path as the bid.

4.6.1   Execution Path Enumeration

Given the state machine of a device and the current state, there are potentially infinite number of paths from the start state. However, we are interested only in those paths where computation of the task can be completed within the advertised deadline. Since deadline is finite, such paths (excluding cycles involving states which offer 0% computing capacity) are also finite in number. Let us denote this set of paths as $\Pi_i = \{\pi_1^i, \pi_2^i, \ldots \pi_k^i\}$ for the device D_i, where k denotes the number of such deadline-constrained paths. A path π_j^i is a state transition sequence, $< s_1^i, s_2^i, \ldots s_m^i >$, in the underlying PFSM-AP of D_i. We assume transitions to be 0-delay.

For each path $\pi_j^i =< s_1^i, s_2^i, \ldots s_m^i >$, we compute the following attributes which are useful for our algorithm.

- *Path execution time* $(\delta^i(\pi_j^i))$: The execution time of the application on the path π_j^i

$$\delta^i(\pi_j^i) = Z + \frac{w_i - Z}{C(s_m^i)} \quad \text{where, } Z = \sum_{l=1}^{m-1} \left(H(s_l^i) \times C(s_l^i)\right)$$

The value of Z denotes the execution time on the first $m - 1$ states on the path. The other term in $\delta^i(\pi_j^i)$ is the time required to finish the remaining fraction of work in the last state (s_m^i).
- *Confidence Value* $(\rho^i(\pi_j^i))$: The confidence value on the path π_j^i is computed as a product of the likelihood values on the transitions (assuming transition probabilities to be independent for simplicity) as below:

$$\rho^i(\pi_j^i) = \prod_{l=2}^{m} \lambda_i(s_{l-1}, s_l)$$

The following constraints are to be applied on valid paths to bound the search.

- *Task completion constraint*:

$$\textbf{C1:} \quad \sum_{l=1}^{m} C(s_l^i) \times H(s_l) \geq w_i$$

Algorithm 3: Bid computation on a mobile device

---

input : J : The task to be executed along with dataset
input : Δ : Deadline for the tasks
input : s : Present state of the device D_i
begin

1	Compute w_i of J;
2	**if** $w_i > \Delta$ **then** No bid and return
3	$b_t \leftarrow \varnothing$ // best time
4	$\Pi_i \leftarrow$ paths (π_j^i) on $\mathcal{U}_i$ satisfying C1 and C2;
5	**for** *each path* $p_j^i \in \Pi_i$ **do**
	if $\rho^i(p_j^i) < \Upsilon$ **then** continue $t \leftarrow \delta^i(p_j^i)$;
	if $(b_t > t)$ **then** $b_t \leftarrow t$
6	Bid with b_t;

---

where the term $\mathbf{C}(s_l^i) \times \mathbf{H}(s_l)$ denotes the quantum of computation done at the state s_l^i considering the average permanence and the CPU availability. The summation on the left-hand side yields the total computation time on a given path. So the above constraint essentially limits our computation to paths whose time is more than w_i.

- *Deadline constraints*: Paths where the completion time of the task is more than Δ are not useful for bidding. Therefore,

$$\mathbf{C2}: \quad \delta^i(\pi_j^i) < \Delta$$

We modify the standard depth-first traversal [29] algorithm with constraints **C1** and **C2**, and also ignore self-loops involving a state with 0% CPU availability. These conditions bound the length of the paths (step 4 of Algorithm 3) to finite values since w_i is finite. Therefore, the algorithm terminates in finite time. The paths enumerated in the state graph are associated with different confidence values. A path with low confidence of traversing should be excluded to avoid penalties. An example of such a computation was presented in Sect. 4.5.1. Algorithm 3 uses a threshold Υ to filter out such paths.

4.7 Experiments with the Offloading System

To evaluate the proposed bidding-based offloading scheme, we built a small-scale cloud computing framework with an *off-loader* residing on a server and distributing jobs to a pool of mobile devices. We present some related results from the system. But, first we present our experiments with a simulation system.

We developed a simulation system to observe the behavior of our task offloading infrastructure. Experiments with our proposed job-offloading technique were carried out with the set of seven mobile devices described in Table 4.1. We simulated the *off-loader* system and also the task execution on the device VMs. Each VM simulates usage of the corresponding device by simulating the logged CPU loads for the last day in the trace file as discussed in Sect. 4.2.2. The simulated *off-loader* generated tasks of various kinds to be offloaded to these devices. When a task is awarded to a device, the winner device simulates the execution of the task while simulating the CPU load replayed from the trace file. For each task type, 100 similar tasks were generated and offloaded to devices based on bids. The number of tasks successfully completed on the devices (i.e., completed within the given deadline) is recorded and used for computing performance of the offloading method. The performance of the system is simply the fraction of the offloaded jobs successfully completed by the bidding device.

In our simulation system, the *off-loader* generates jobs of various durations, assigns various deadlines to these jobs, invites bid, and submits the job to the winning device. Our job-offloading experiment was conducted for jobs whose execution time ranges from 4 to 29, and deadlines varying from 2× to 10× of the job execution time. The offloading performance, as discussed earlier, is the fraction of the number of jobs the infrastructure could complete by offloading them to winning devices and the devices subsequently could complete execution within the given deadline. The result of the simulation experiment is shown in Fig. 4.6, where the horizontal axis represents variation in job execution duration and the vertical axis represents the offloading performance. A value of 0 as performance indicates that the infrastructure was unable to effectively utilize any device for computation. On the other hand, a value of 1 indicates the infrastructure could execute all jobs using the devices. Please note that, in our experiment, the infrastructure offloaded one job at a time and concurrent offloading was not considered.

It is evident from the experiments that deadline is the most important factor for offloading tasks. When the deadline is very tight in comparison to the execution time of the task, task offloading is not beneficial. When the deadline is tight, if the device cannot operate with near 100% CPU availability all the time, the execution time is

Fig. 4.6 Simulated system performance

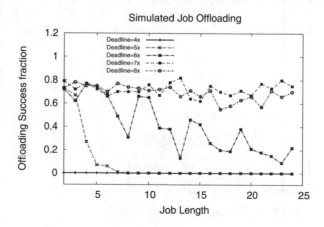

more likely to overshoot the task deadline. When the deadline is very relaxed (e.g., approximately $6\times$ that of job execution length) offloading technique works well and is beneficial. It is also evident from the result that very short tasks with very tight deadlines are not suitable to exploit this mobile computing system, rather a moderately high computation job with a relaxed deadline is more suitable from this framework. Another interesting observation is that for higher deadlines, offloading success does not improve at the same scale.

4.7.1 Working with Real Devices

In this phase, we evaluated the performance of our offloading system with an *off-loader* which has the seven devices, described earlier, for it to exploit. The details of the system are described below.

4.7.2 Architecture of Offloading System

The architecture of the system is presented in Fig. 4.7. The *off-loader* is composed of two principle components—(1) *Job Manager* which is responsible for managing incoming tasks. In our system setup, this module generates tasks to be scheduled in mobile devices. (2) *Bid Manager* is responsible to query available devices and then initiate bidding for each task, obtained from the job manager. The bid manager

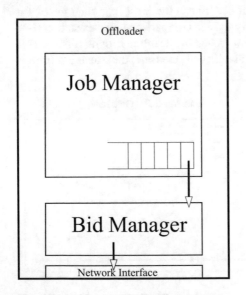
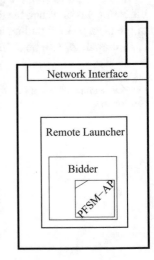

Fig. 4.7 Architecture of offloading system

Fig. 4.8 Experimental
system performance

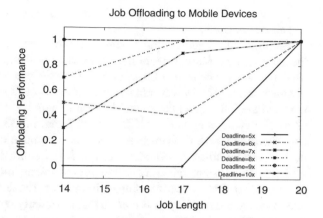

creates a socket connected to a pre-defined IP of the server machine. Each of
the mobile devices which volunteer to donate their CPU cycles need to install a
lightweight android application, called *Remote Launcher*. The application connects
and communicates with the *bid manager* of the *off-loader* through the pre-defined
socket. The *bid manager* advertises a task on all the connections. The *remote
launcher* computes its bid and sends back as a response to the *bid manager*. On
award of a job, the *remote launcher* forks the job in local mobile devices, collects
the result, and sends the data back to the *bid manager*. All the devices communicate
using the WiFi network provided in our lab and the devices always remain within
the communication range during the experiment.

For this experiment, we used an application which estimates the value of π,
which is written as a native android application. The application is a compute-
intensive one. Longer the application runs, the estimation is better. We conducted the
experiment for various job durations and the job duration was varied by changing
the desired accuracy of π-value calculation. Figure 4.8 shows the job-offloading
performance of the system. The result of this experiment shows that jobs with
relaxed deadlines are good candidates for offloading.

4.8   Related Work

In recent times, there has been a significant volume of research on the theme
of mobile cloud computing (MCC), which proposes the use of a collaborative
computing infrastructure consisting of mobile devices and a back-end cloud computing system. MAUI [22] and CloneCloud [23] are the two notable systems which
use a back-end computing infrastructure for collaborative job execution in mobile
devices. Several other articles as well address the problem of work-flow partitioning
in an MCC setting [30]. The basic intuition behind these partitioning strategies is to
decide on the best platform (mobile device or cloud) to execute each sub-task of a

given work-flow with an objective of optimizing the cost (in terms of energy, time, and communication).

Authors in [31] address the problem of intermittent disconnection and analyze the same using a Markov-chain model. Markov decision process (MDP) has been used to model the behavior of mobile devices to achieve objectives like optimization of power usage [32]. However in our case, we resort to a simpler model since we do not need the full capabilities of an MDP for this problem. A comprehensive survey of mobile cloud computing (MCC) can be found in [33]. Systems like *Misco* [7] and *Hyrax* [6] extend Map-Reduce so that computation capabilities of mobile devices can be utilized. *Serendipity* is a task dissemination system over a mobile grid [8]. The system relies on collaboration among mobile devices using WiFi connection to share collective computation power. The design of the system accepts that disconnection of devices is a norm and its underlying architecture incorporates the assumption. We consider a more generic usage model driven scenario in this work. Usage model-based task scheduling on mobile devices is attracting more attention in recent years. Authors in [3] present an analysis of the usage patterns of the communication module of a device. They define a state space composed of five states in which a device operates and present their analysis based on transitions among these states. Authors in [34] present a framework which analyzes pattern of usage of an application by its user and suggest a mechanism for optimizing launch time of the application. They claim that application launch latency is improved by this method which is reflected in improvement in user experience with the device.

4.9 Conclusion and Future Work

In this paper, we present a state transition-based system to model the usage pattern of a mobile device. The model captures the variation of free resources in the device based on its owner's usage pattern. We used this model to schedule background tasks in the device such that usage experience of the device does not degrade due to consumption of resources by the background tasks. We also used the same model in a different scenario of mobile cloud collaborative computing. We present a system, based on bidding, where mobile devices perform some tasks which a cloud computing infrastructure tends to offload to a set of participating mobile devices. Our usage pattern model, although being simple, can be effectively used in such diverse scenarios. Mechanisms for automatic learning of likelihood probabilities, using more advanced models for analysis, execution time estimation, designing more effective bidding, and reward-penalty schemes may be looked into for future explorations. In this paper, we also assume the network and the devices are reliable. Issues of fault tolerance in this context are our future research agenda.

References

1. T. Danova, Gartner: mobile apps will have generated $77 billion in revenue by 2017 (2014), http://e.businessinsider.com/public/2373445
2. S. Sakr, Nvidia says Tegra-3 is a "PC-class CPU" (2011), http://engt.co/srvibU
3. J.-M. Kang, S.-s. Seo, J.-K. Hong, Usage pattern analysis of smartphones, in *2011 13th Asia-Pacific Network Operations and Management Symposium (APNOMS)* (IEEE, Piscataway, 2011), pp. 1–8
4. J.-M. Kang, S.-s. Seo, J.W.-K. Hong, Personalized battery lifetime prediction for mobile devices based on usage patterns. J. Comput. Sci. Eng. (4), 338–345 (2011)
5. A. Shye, B. Scholbrock, G. Memik, P.A. Dinda, Characterizing and modeling user activity on smartphones: summary, in *ACM SIGMETRICS Performance Evaluation Review*, vol. 38, no. 1 (ACM, New York, 2010), pp. 375–376
6. E.E. Marinelli, Hyrax: cloud computing on mobile devices using MapReduce. Carnegie-Mellon Univ, School of Computer Science, Pittsburgh, PA, Tech. Rep. CMU-CS-09-164, Sept 2009
7. A. Dou, V. Kalogeraki, D. Gunopulos, T. Mielikainen, V.H. Tuulos, Misco: a MapReduce framework for mobile systems, in *Proceedings of the 3rd International Conference on Pervasive Technologies Related to Assistive Environments* (ACM, New York, 2010), p. 32
8. C. Shi, V. Lakafosis, M.H. Ammar, E.W. Zegura, Serendipity: enabling remote computing among intermittently connected mobile devices, in *Proceedings of the Thirteenth ACM International Symposium on Mobile Ad Hoc Networking and Computing* (ACM, New York, 2012), pp. 145–154
9. A. Carroll, G. Heiser, An analysis of power consumption in a smartphone, in *USENIX Annual Technical Conference*, pp. 1–14 (2010)
10. L. Ardito, G. Procaccianti, M. Torchiano, G. Migliore, Profiling power consumption on mobile devices, in *ENERGY 2013, The 3rd International Conference on Smart Grids, Green Communications and IT Energy-Aware Technologies*, pp. 101–106 (2013)
11. A. Schulman, V. Navda, R. Ramjee, N. Spring, P. Deshpande, C. Grunewald, K. Jain, V.N. Padmanabhan, Bartendr: a practical approach to energy-aware cellular data scheduling, in *Proceedings of the 16th Annual International Conference on Mobile Computing & Networking* (ACM, New York, 2010), pp. 85–96
12. A. Chakraborty, V. Navda, V.N. Padmanabhan, R. Ramjee, Coordinating cellular background transfers using loadsense, in *Proceedings of the 19th Annual International Conference on Mobile Computing & Networking* (ACM, New York, 2013), pp. 63–74
13. P.K. Athivarapu, R. Bhagwan, S. Guha, V. Navda, R. Ramjee, D. Arora, V.N. Padmanabhan, G. Varghese, Radiojockey: mining program execution to optimize cellular radio usage, in *Proceedings of the 18th Annual International Conference on Mobile Computing & Networking* (ACM, New York, 2012), pp. 101–112
14. A. Shye, B. Scholbrock, G. Memik, Into the wild: studying real user activity patterns to guide power optimizations for mobile architectures, in *Proceedings of the 42nd Annual IEEE/ACM International Symposium on Microarchitecture* (ACM, New York, 2009), pp. 168–178
15. V. Tiwari, S. Malik, A. Wolfe, M.T.-C. Lee, Instruction level power analysis and optimization of software, in *Technologies for Wireless Computing* (Springer, New York, 1996), pp. 139–154
16. S. Hao, D. Li, W.G. Halfond, R. Govindan, Estimating mobile application energy consumption using program analysis, in *2013 35th International Conference Software Engineering (ICSE)* (IEEE, Piscataway, 2013), pp. 92–101
17. R. Wilhelm, J. Engblom, A. Ermedahl, N. Holsti, S. Thesing, D. Whalley, G. Bernat, C. Ferdinand, R. Heckmann, T. Mitra et al., The worst-case execution-time problem - overview of methods and survey of tools. ACM Trans. Embed. Comput. Syst. (TECS) **7**(3), 36 (2008)
18. M.R. Garey, D.S. Johnson, R. Sethi, The complexity of flowshop and jobshop scheduling. Math. Oper. Res. **1**(2), 117–129 (1976)

19. D.G. Feitelson, Job scheduling in multiprogrammed parallel systems: extended version, IBM research RPT. RC **19790**, 87657 (1979)
20. F. Bonomi, R. Milito, J. Zhu, S. Addepalli, Fog computing and its role in the internet of things, in *Proceedings of the First Edition of the MCC Workshop on Mobile Cloud Computing* (ACM, New York, 2012), pp. 13–16
21. A. Mukherjee, H.S. Paul, S. Dey, A. Banerjee, Angels for distributed analytics in IOT, in *2014 IEEE World Forum on Internet of Things (WF-IoT)* (IEEE, Piscataway, 2014), pp. 565–570
22. E. Cuervo, A. Balasubramanian, D.-k. Cho, A. Wolman, S. Saroiu, R. Chandra, P. Bahl, MAUI: making smartphones last longer with code offload, in *Proceedings of the 8th International Conference on Mobile Systems, Applications, and Services* (ACM, New York, 2010), pp. 49–62
23. B.-G. Chun, S. Ihm, P. Maniatis, M. Naik, A. Patti, CloneCloud: elastic execution between mobile device and cloud, in *Proceedings of the Sixth Conference on Computer Systems* (ACM, New York, 2011), pp. 301–314
24. S. Kosta, A. Aucinas, P. Hui, R. Mortier, X. Zhang, ThinkAir: dynamic resource allocation and parallel execution in the cloud for mobile code offloading, in *INFOCOM, 2012 Proceedings IEEE* (IEEE, Piscataway, 2012), pp. 945–953
25. M.S. Gordon, D.A. Jamshidi, S. Mahlke, Z.M. Mao, X. Chen, Comet: code offload by migrating execution transparently, in *Proceedings of the 10th USENIX conference on Operating Systems Design and Implementation, OSDI*, vol. 12, pp. 93–106 (2012)
26. T. Phan, L. Huang, C. Dulan, Challenge: integrating mobile wireless devices into the computational grid, in *MobiCom '02: Proceedings of the 8th Annual International Conference on Mobile Computing and Networking, MOBICOM-2002*, pp. 271–278 (2002)
27. A. Agarwal, Enterprise smartphone usage trends (2011), http://bit.ly/loIqE1
28. X. Li, A. Gray, D. Jiang, X. Mao, Sufficient and necessary conditions of stochastic permanence and extinction for stochastic logistic populations under regime switching. J. Math. Anal. Appl. **376**(1), 11–28 (2011)
29. R. Tarjan, Depth-first search and linear graph algorithms. SIAM J. Comput. **1**(2), 146–160 (1972)
30. M.R. Rahimi, N. Venkatasubramanian, A.V. Vasilakos, MuSIC: mobility-aware optimal service allocation in mobile cloud computing, in *Proceedings of the 2013 IEEE Sixth International Conference on Cloud Computing, CLOUD '13* (IEEE Computer Society, Washington, DC, 2013), pp. 75–82
31. S.-M. Park, Y.-B. Ko, J.-H. Kim, Disconnected operation service in mobile grid computing, in *Service-Oriented Computing-ICSOC 2003* (Springer, New York, 2003), pp. 499–513
32. E. Jung, F. Maker, T.L. Cheung, X. Liu, V. Akella, Markov decision process (MDP) framework for software power optimization using call profiles on mobile phones. Design Autom. Embed. Syst. **14**(2), 131–159 (2010)
33. H.T. Dinh, C. Lee, D. Niyato, P. Wang, A survey of mobile cloud computing: architecture, applications, and approaches. Wirel. Commun. Mob. Comput. (2011)
34. W. Song, Y. Kim, H. Kim, J. Lim, J. Kim, Personalized optimization for android smartphones. ACM Trans. Embed. Comput. Syst. (TECS) **13**(2s), 60 (2014)

Chapter 5
Cyber-physical Autonomous Vehicular System (CAVS): A MAC Layer Perspective

Jahan Ali, Md Arafatur Rahman, Md Zakirul Alam Bhuiyan, A. Taufiq Asyhari, and Mohammad Nomani Kabir

5.1 Introduction

In vehicular networking for intelligent transportation, cyber-physical autonomous vehicular system (CAVS) is a promising technology that has been widely considered by governments, autonomous industries, and academic research institutes [60, 73]. CAVS is the synergistic integration of heterogeneous networking, computation, and physical processes in which all vehicles and their components communicate via a vehicular networking platform and are driven in a platoon-based pattern with a closed feedback loop between the cyber process and physical process [63]. The main purpose of the vehicular cyber-physical systems is to incorporate computation, communication, and control to enhance road safety, efficiency, convenience, and high quality of daily life by minimizing traffic congestion and injuries, and improving fuel efficiency while traveling on the road [51].

J. Ali · M. N. Kabir
Faculty of Computer Science and Software Engineering, University Malaysia Pahang, Pahang, Malaysia

Md. A. Rahman
Faculty of Computer Science and Software Engineering, University Malaysia Pahang, Pahang, Malaysia

IBM Center of Excellence, University Malaysia Pahang, Pahang, Malaysia

M. Z. A. Bhuiyan (✉)
Department of Computer and Information Sciences, Fordham University, Bronx, NY, USA

A. T. Asyhari
Centre for Electronic Warfare, Information and Cyber, Cranfield University, Defence Academy of the UK, Shrivenham, UK

© Springer Nature Switzerland AG 2020
S. Hu, B. Yu (eds.), *Big Data Analytics for Cyber-Physical Systems*,
https://doi.org/10.1007/978-3-030-43494-6_5

129

Furthermore, with the recent advancement of automation in automotive industries, the vehicular wireless communication becomes an emerging research interest both for automotive industry engineers and researchers [13, 111]. As a consequence, there is an increasing number of required safety features in the modern vehicle such as driver safety, road safety, vehicle to infrastructure recognition, and vehicle-to-vehicle communications. The current generation of vehicles integrates a number of various sensors with the help of different wireless communication protocols to provide real-time communication for those purposes. Incorporating a large number of sensors inside the vehicle provides overall safely support for vehicle as well driver, whereas designing an interoperable, reliable, and flexible communication technology has been the key challenge [83, 116].

A wireless sensor network (WSN) is spatially characterized by various autonomous sensors to supervise physical or environmental conditions, such as remote health monitoring, temperature, sound, pressure, etc. and to simultaneously forward their particular information over a communication network to a sink location [17, 18, 46]. The combination of WSN with internet of things (IoT) enables WSN with either object-oriented or internet-centric resources, which enormously enlarges the capacity of WSN in a large number of sensing and communication applications [53, 55].

Modern smart automobiles are monitored by complex distributed networks made up of a large number of heterogeneous wireless sensor nodes with rich connectivity supplied by central networks and internet [65]. With the rapid improvement of the automobile intelligent system and intelligence and internet connectivity, security and privacy have emerged to be the main challenges for automotive systems [98]. Researchers have shown that by reducing a single control unit, a possible attacker may obtain access to other vehicle controller units via internal communication buses such as controller area network (CAN), and harm critical subsystems [111]. As CAN receives simultaneous connection with IoT resources and service providers, it becomes effortless marks to cyber adversaries, particularly, because, it has never been supposed to handle cyber risks. This makes CAN information susceptible to falsehood assaults that turn into erroneous critical information distribution to users, which in turn leads the system to take improper and hazardous activities or to be unaware of an ongoing attack as was the case in Stuxnet attack [80, 118]. It also enables adversaries to possibly perform destructive instructions on control systems, causing dangerous actions (e.g., disabling the brake system). Therefore, it is crucially important to secure and protect smart automobile functions towards any kinds of cyber-related assaults [80, 117].

CAVS is an integration of physical process and cyber systems via heterogeneous networking and communications. A typical CAVS is shown in Fig. 5.1, comprising physical components (such as vehicles, DSRC/mobile devices, tablets), cyber systems (e.g., data center, traffic control center), and communications (e.g., vehicular networks).

The CAVS industry has noticed a progressive improvement since 2014 and is estimated to continually exhibit a significant growth as shown in Fig. 5.2 [93]. The concept of the CAVS is growing in every industry sector and more and

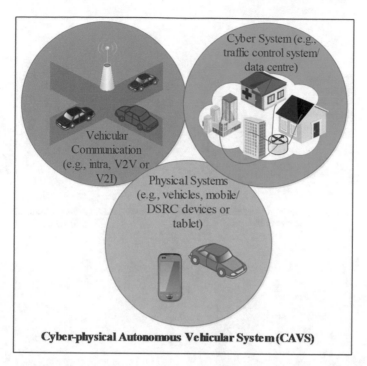

Fig. 5.1 A typical diagram illustrating the concept of vehicular cyber-physical systems using the interactions among system components

more wireless technologies are being incorporated in the emerging IoT protocol stack [28]. Device-to-Device (D2D) communication is innovating in daily basis and envisioned to create a wireless ecosystem of billions of intelligent electronics devices within a single entity named IoT [15]. In the CAVS ecosystem, intelligent devices communicate with each other autonomously to assemble, communicate, and forward heterogeneous information in a multi-hop manner without any human centralized control and collaboration. However, the real-time communication among the intelligent devices is the key leveraging value in CAVS intelligent environment where information is gathered and transformed intelligently. Ultimately, the ability to gather different types of information varies from one device to another, which is particularly driven by assorted networking standards and connectivity challenges [16, 62].

5.2 Impact of CPS on Vehicular Systems

Various automobile manufacturers are spending investment into the cyber-physical system (CPS), and one particular concern is CPS connected automated vehicles.

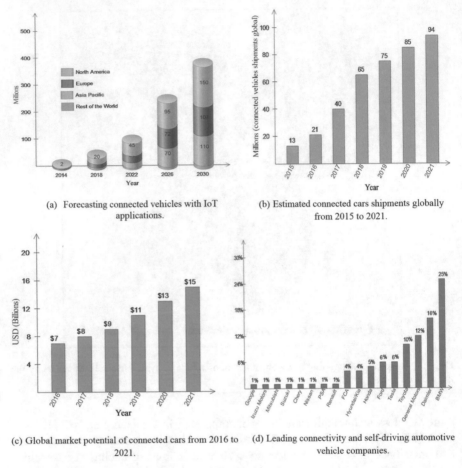

(a) Forecasting connected vehicles with IoT applications.

(b) Estimated connected cars shipments globally from 2015 to 2021.

(c) Global market potential of connected cars from 2016 to 2021.

(d) Leading connectivity and self-driving automotive vehicle companies.

Fig. 5.2 Estimating and forecasting connecting vehicles due to the emerging growth of CAVS

Business intelligence (BI), a business insider's premium research service, anticipated 94 million connected automobiles to ship in 2021 as shown in Fig. 5.2b, which constitute to 82% of all vehicles transported in the same year. This would represent a compound yearly growth rate of 35% from 21 million connected vehicles in 2016. Automobile marketplaces have successfully observed a flourishing trend and a significant business opportunity for their connected automobiles. BI forecasted 381 million connected cars to be on the road by 2020, up from 36 million in 2015. Furthermore, BI forecasted that this will generate $8.1 trillion from connected automobiles between 2015 and 2020 as shown in Fig. 5.2c.

Automobile manufacturers are becoming interested in heavily investing in their connected automobile initiatives due to a number of reasons. Internet connectivity in vehicles enables automobile manufacturers to introduce software updates in real time, which are extremely worthwhile for after sales service. Moreover, the automobile manufacturers can leverage upon a huge amount of big data from vehicles to analyze the performance and generate invaluable insights on how drivers utilize their vehicles. Furthermore, the vehicle connectivity provides more efficient ways to sell their products and better quality of services to customers. It is expected that in the near future the connected vehicles in the CPS ecosystem produce a big amount of data which will pave the way to ensure the driving safety of fully reliable CAVS via vehicle-to-vehicle (V2V) and vehicle-to-infrastructure (V2I) communications.

Recently, many leading automotive companies have introduced the connected vehicle concept in the CPS ecosystem as shown in Fig. 5.2d according to the KPMG survey of 200 automotive executives. The BMW is the top leading company with other manufacturers including Daimler, General Motors, Toyota, and Tesla being near the top of the list.

The CPS can be enabled with the advanced of development in wireless technologies, including RFID, smart sensors, communication technologies, and internet protocols. Basically, the main principle is driven by the need of smart sensors to cooperate directly without any human interaction. The recent change in internet-centric phenomena, mobile communications, and overall machine-to-machine (M2M) communication technologies can be seen as the fundamentals of the CPS [4, 81]. In the coming years, the growth of the CPS is predicted through integration of enormous wireless communication technologies to enable unique applications by connecting cyber systems with physical objects together in support of intelligent decision making [105]. Likewise, today's automobiles have an increased range of effective electronic control devices (ECU) and associated distributed sensor and actuator elements [68]. For instance, a significant number of more than 50 sensors are implemented nowadays in a mid-range vehicle, while the industry estimates for any automotive sensor market volumes surpass 665 million devices and 2237 million devices, in the USA and globally, respectively [35]. This represents a multi-billion dollar market in electronic sensors alone in 2007 and the analysts estimated a significantly more than 80% of the latest functions in cars become electronic devices based [3, 84, 85]. As a result, the designing concept, production and installation of the wiring for several of these sensor elements require considerable engineering work. At present, wiring harnessed inside a vehicle with 4000 parts may have weighed up to 40 kg and up to 4 kilometers long [65]. Eliminating or reducing the number of wires may potentially provide mass savings, warranty savings, and overall cost savings. This opportunity also paves the way to enable wireless technology inside the vehicle to monitor its critical parts as envisioned in the concept of intelligent transportation system (ITS). Therefore, the intra-vehicular wireless communication requires an in-depth investigation of various types of wireless communication protocols for integration within the CAVS [83].

Modern automobiles are progressively furnished with a large number of different types of sensors, actuators, and communication protocols and devices (WSNs, 3G/4G, mobile devices, GPS devices, and embedded computers) [66]. Generally, automobiles have been equipped with effective sensing, networking, communication, and data processing capabilities, and can communicate with other automobiles or transfer important information with the external environments over various internet protocols like HTTP or TCP/IP, and next-generation telematics protocols [107]. Consequently, various advanced telematics solutions such as remote security for disabling the engine and remote diagnosis have been designed to improve drivers' safety, efficiency, and enjoyment [30]. The improvements in cloud computing and CAVS have delivered a promising opportunity to further address the increasing intelligent transportation issues, such as heavy traffic, congestion, and vehicle safety [41]. In the recent years, researchers have suggested a few models that use cloud computing for implementing ITSs. For example, a new vehicular cloud architecture called ITS-Cloud was proposed to improve V2V communications and road safety [45].

The integration of sensor devices and various both wired and wireless communication technologies pave the way for us to track the updating status of an object by using the internet. The CPS describes a future in which a number of physical objects and devices around us, such as various sensors, radio frequency identification (RFID) tags, GPS devices, and mobile devices, will be associated to the internet and allows these objects and devices to connect, cooperate, and communicate within social, environmental, and user contexts to reach common goals [20, 41, 125]. As a promising technology, the CAVS is expected to offer appealing possibilities to transform transportation systems and automobile services in the automobile industry [38]. Guerrero-ibanez et al. [41] proposed an idea to use the "unique identifying properties of car registration plates" to connect various things. As vehicles are equipped with increasingly powerful sensing, networking, communication, and data processing capabilities, CAVS technologies can be used to harness these capabilities and share under-utilized resources among vehicles in the parking space or on the road. For example, CAVS technologies can make it possible to track each vehicle's existing location, monitor its movement, and predict its future location [52].

By integrating cloud computing with WSNs, RFID tags, satellite network, and other ITSs technologies, a new generation of CPS-based vehicular data clouds can be improved to bring many business advantages, such as anticipating improved road safety, decreasing road traffic congestion, controlling traffic, and promoting car maintenance or repair [26]. Some exploratory works of using CPS technologies to enhance ITSs have been performed in recent few years [2, 115]. For example, an intelligent informatics system (iDrive system) manufactured by BMW used various sensors and tags to monitor the environment, such as tracking the vehicle location and the road condition, to provide driving directions [8, 45]. Anand et al. and Uden et al. [8, 113] proposed an intelligent internet-of-vehicles system (known as IIOVMS) to collect traffic information from the external environments on an ongoing basis in order to monitor and manage road traffic in real time. He et al. [45] discussed how ITSs could use IoT devices in the vehicle to connect to the cloud and how numerous sensors on the road could be virtualized to leverage the processing capabilities of the cloud. Anand et al. [8] proposed a technology architecture that uses cloud computing, IoT, and middleware technologies to enable innovation of automobile services [123].

WSNs are mainly characterized by the limited resources of the nodes. A holistic network configuration and planning is crucial for the effective uses of the restricted resources. The media access control (MAC) protocol, as a fundamental part of the networking stack, should be configured with respect to the topological structure of the network, the power source of the nodes, and the characteristics and requirements of the running applications [14, 70, 120].

In the subsequent section of this article, some general challenges of MAC protocol development for CAVS are first reviewed. It follows with state-of-the-art MAC protocol concept for the CAVS platform. A proposed framework is then suggested for the future CAVS-enabled vehicular WSN. A conclusion is drawn to highlight the key points discussed in this article.

5.3 Challenges of Developing Reliable MAC Strategies in CAVS

A wide range of issues need to be addressed at different layers of the architecture and from different aspects of system design to improve the CAVS. In this section, some general challenges of wireless MAC design in the CAVS are discussed:

Large-Scale Deployment and Ad Hoc Architecture Most of the networks contain a large number of deployed sensor nodes without a predetermined network infrastructure, which exhibits challenges to provide seamless autonomous connectivity. Developing a suitable technology to design a CAVS from existing technologies requires a comprehensive analysis on communication protocols of the selected technology. In addition, new technology can also be explored for such a network.

Integration with Internet and Other Networks It is a fundamental importance for the CAVS to provide continuous services that allow querying of the network to retrieve useful information from anywhere and at any time. Therefore, the vehicular wireless technology should be remotely accessible from the internet and needs to be integrated with the existing internet protocols.

Resource-Efficient MAC An energy-efficient protocol for wireless communication is important to optimize the network lifetime of the CAVS. The energy saving can be achieved in every feature of the network by combining network functionalities with energy-efficient protocols such as energy-aware routing on network layer and energy-saving mode on MAC layer.

Self-Configuration and Self-Organization MAC The MAC of the existing technologies should have dynamic topology features to avoid the node failure due to mobility and large-scale mobile node deployments that demand self-organizing architectures and protocols. New sensor nodes can be incorporated to replace the failing sensor nodes in the implementation area and, similarly, existing nodes in the network can also be eliminated from the system without impacting the common objective of the application.

Quality of Services As much as possible, the CAVS connection is required to be safe and reliable without minimal human interaction. The QoS provided by the MAC layer corresponds to the accuracy between the data reported to the sink node and what is actually occurring in the physical environment. In the CAVS, it is undesirable to have sensor data with long latency due to processing or communication because they may be outdated and lead to wrong decisions in the monitoring system.

5.4 State-of-the-Art on MAC Design and Development towards Reliable CAVS

Different of technologies to enable the IoT are becoming widely available due to the need to provide a better understanding of our environment [77]. As a result, intelligent devices and networks embedded in various kinds of WSNs are connected and integrated with the IP-based large network [1, 5]. However, many open challenges, mostly the suited protocols and standards, still persist in practical implementation. Specifically, the MAC layer carries a fundamental building block of WSNs to establish the communication link among different network infrastructures [87, 101]. Several relevant classifications of IoT enabled MAC protocols in WSNs have been presented based on operating principles and underlying features to emphasize their opportunities, strengths, and weaknesses [59].

Within a pervasive sensing framework, WSN technologies have been an integral aspect of IoT, which enable sensors and actuators to communicate seamlessly with the environment around us, and to share the information across multiple platforms towards the vision of a smart environment system [22, 56, 114]. Smart connectivity with existing network infrastructures and ubiquitous computation using network resources is an indispensable part of IoT. However, the success of IoT depends on the improvement of the network performance, flexibility, interoperability, reliability, and limited energy consumption of WSNs [106]. Among various network layers, the MAC layer protocols have been received more attention for development towards reliable energy-efficient sensor network architecture. The attributes of the MAC layer significantly impact on the performance, power consumption, and scalability of the sensor network [89].

Since the past few years different wireless sensor MAC protocols have been successfully developed to fulfill the requirement of the growing scale of the WSNs in the CAVS [29, 57, 89]. This article classifies these MAC protocols according to different channel access mechanisms, which include resource sharing methods, contention based, channel polling, scheduling based, and hybrid MAC protocols. As an overview, the main classification of the MAC protocols is given in Fig. 5.3.

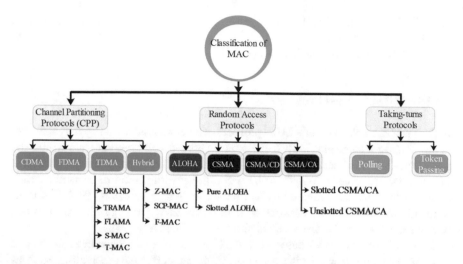

Fig. 5.3 Classification of MAC protocols for CAVS

5.4.1 Channel Partitioning Protocols (CPP)

The resource division method along one or more dimensions for a MAC protocol has been well established. There exist three recognized resource division methods, namely frequency-division multiple access (FDMA), time-division multiple access (TDMA), and code-division multiple access (CDMA). The FDMA strategy splits the resource into several partitions of channels. TDMA mechanism divides the resource into multiple time slots and the CDMA method separates the resource into a collection of codes in which the same channel is identified by the assigned user [43, 49, 86, 88].

TDMA is a synchronous method wherein the users cannot communicate individually and simultaneously. No overlapping time slots are permitted to assorted users to increase the data rate, since it can only access its allocated time slot. TDMA typically adheres to a high power efficiency requirement, which is the most desirable characteristics for low power operating systems, at the expense of reducing the transmission capacity per user [10, 27, 112].

FDMA transmission permits several users to send simultaneously using different frequencies inside a uniform cell per room topology. In a cellular strategy, the data transfer is divided into non-overlapping frequency bands. In orthogonal frequency-division multiple access (OFDMA), users are allocated frequency slots spanning several OFDM symbols and subcarriers. In general, power efficiency is the main disadvantage of FDMA and it worsens as the number of subcarriers increases [32, 79, 89].

For CDMA techniques, the typically used strategies rely on direct sequence distribution where users can receive the exact channel using optical orthogonal codes (OOCs). This corresponds adaptability of adding users and asynchronous accessibility capacity [64]. Users are allowed to transmit at overlapping times and wavelengths. Consequently, it is feasible to implement hybrid systems such as WDMA/CDMA or TDMA/CDMA [44, 67].

5.4.1.1   TDMA Based Protocols

In the TDMA mechanism of a MAC protocol, a channel is accessed by only one user at any time in a same slot [44]. If the other users want to communicate within the same time slot, it will lead to packet losses or network collision. The main drawback of using this scheme in WSNs is the increasing waiting time to transmit, which increases the network vulnerability and reduces network scalability [91]. However, to avoid any data drop or network collision, this MAC strategy is recommended because it can utilize the network resource more effectively [74].

Several MAC strategies based on the TDMA mechanism have been developed over the last few years, which focus on different performances metrics such as energy efficiency and quality of service (QoS), and result in collision-free MAC protocols such as TRAMA [58, 72, 104]. The S-MAC and T-MAC follow the synchronous approach schemes, which tolerate a little schedule misalignment in the network, although they still require a globally synchronized schedule that creates an additional energy overhead [71]. S-MAC is popular as an energy-efficient protocol, but still fails to guarantee to a satisfactory QoS performance in the large-scale network topology [121].

S-MAC utilizes a mixed contention and scheduling scheme for collision avoidance. Furthermore, interfering nodes will go to sleep when they receive control message to avoid overhearing [34]. In S-MAC, lengthy messages will be separated into tiny fragments, which will be sent as a burst. It creates additional messages to send, which need a longer accessibility to the medium. S-MAC is developed primarily to reduce energy consumption at the expense of sacrificing other important performance factors such as fairness, throughput, bandwidth utilization, and latency [33, 61]. Fairness will degrade (from a MAC level perspective) as some nodes with small time frame will require to wait MAC access with adaptive listening in which messages shift two hops in every duty cycle. Consequently, latency becomes higher as more messages are prepared to be sent [24].

T-MAC is designed to achieve a better performance over S-MAC by utilizing a dynamic duty cycle instead of a fixed one [6]. The concept is to transfer all information from one node to another in bursts of adjustable length and introduce sleeping between bursts for additional energy efficiency [40]. It can also decide the duration of the variable load by keeping an optimal time. T-MAC pertains RTS and CTS methods. When RTS cannot obtain a CTS response, it would attempt again before giving up [48]. As in S-MAC, T-MAC can only deliver the information to one single hope for each duty cycle, which results in a large latency [9]. Additionally, T-MAC has an early sleep problem as a node changes to sleep even when a neighbor has some information waiting to be sent. Consequently, the throughput is reduced at the sink nodes [124]. T-MAC can adjust the duty cycle in accordance with the traffic load of the network. It also offers scalability and collision avoidance functionality through a mixed scheduling and contention schemes such as S-MAC [37]. Table 5.1 shows the key principle differences among S-MAC, T-MAC, and DSMAC.

Several wireless MAC protocols that overcome the difficulties of receiving global topology information in the large and scalable networks such as DRAND, PACT, and TRAMA have been proposed in the literature [7, 95, 102]. Flow-aware medium access (FLAMA) is a TDMA based MAC protocol modified from TRAMA, which is prominent for periodic monitoring applications [58]. The main idea of FLAMA is to prevent the overhead corresponding to the exchange of traffic information. As the information movement in periodic reporting applications is rather stable, FLAMA first sets up the flows and then uses a pull-based mechanism so that data is transferred only after being explicitly requested [39, 75].

5.4.1.2 FDMA Based Protocols

FDMA is another strategy that provides a collision-free channel. In practice, FDMA needs additional circuitry to dynamically communicate with various radio channels [47]. This operation increases the cost of the sensor nodes, which contradicts with the objective of sensor network systems. Compared to TDMA and CDMA, FDMA is less suitable for operation in low-cost devices [37]. The underlying reason for this is that FDMA capable nodes need extra circuitry to communicate over and change among different radio channels. The complicated band pass filters needed for this operation are reasonably expensive. Another drawback of FDMA that restricts its practical use is the rather strict linearity requirement on the medium [11].

5.4.1.3 CDMA Based Protocols

CDMA also provides a collision-free medium access mechanism. Its main characteristic is the high computational requirement, which is a major barrier for the required energy-efficient sensor networks [82]. To minimize the computational time in the CDMA based wireless sensor networks, there has been limited effort to investigate the computationally feasible source and modulation schemes, particularly signature waveforms, simple receiver models, and other signal synchronization schemes [27]. If it can be shown that the high computational complexity of CDMA can be traded-off with the collision avoidance function, then CDMA protocols might also be regarded as a possible solution for sensor networks [32].

5.4.1.4 Hybrid Protocols

Hybrid MAC protocols are usually a combination of TDMA, FDMA, and CDMA. Protocols that combine TDMA and CDMA such as reservation-based and contention-based hybrid MAC protocols and a TDMA/FDMA based hybrid

Table 5.1 Key principle differences among S-MAC, T-MAC, and DSMAC

References	MAC name	Targeted application	Key design principles	Strengths	Weakness
Huang et al. [47], Rao et al. [90]	S-MAC	Bursty event Multihop	Fixed low duty cycle, Maintain NAV for virtual carrier sensing (virtual clustering) Use physical/virtual carrier sense with randomized carrier sense time, RTS/CTS exchange and NAV to avoid overhearing	Low duty cycle to save energy, virtual clusters to support scalability and self-configuration, overhearing avoidance to save energy, message passing to reduce contention latency	High latency due to periodic sleep, fixed duty cycle not adaptive to dynamic traffic loads
Arifuzzaman et al. [9], Suriyachai et al. [108]	T-MAC	Dynamic traffic loads in time and location, Multihop	Transmit messages in a burst of variable lengths, adaptive duty cycle (ADC) with timeout mechanism dynamically ending the active part, future request-to-send (FRTS), full-buffer priority with threshold control	Save more energy by adaptation to dynamic traffic	ADC increases latency and reduces throughput, difficult to distinguish the communication pattern of a live WS
Doudou et al. [34], Huang et al. [47], Yin et al. [122]	DSMAC	Dynamic traffic loads to meet the application's load	Fully independent duty cycle, so that each node can adapt, in a fully distributed way, to the current surrounding conditions	Energy Efficient, dynamically changes each node's duty cycle	Additional energy overhead

MAC protocol called HYMAC behave like CSMA at low contention levels and switch to TDMA-type operation at high contention levels [109]. Protocols such as the hybrid MAC proposed in [100] combine CSMA with TDMA and FDMA where nodes are allocated a frequency as well as a time frame to send data once they effectively request for bandwidth resources using contention-based transmission. Similar protocols where CSMA-based bandwidth demands are used to decide the assignment of timeslots and codes have also been proposed in [103]. Zebra MAC (Z-MAC) protocol is one of the most widely considered examples in a hybrid scheme, which integrates the strengths of both TDMA and CSMA while offsetting their disadvantages [96, 97]. The Scheduled Channel Polling MAC (SCP-MAC) and Funneling-MAC protocol are other alternative examples of the widely used hybrid MAC schemes [54].

Table 5.2 presents a broad classification of the various hybrid MAC protocols. There are two types of contention-free access scheme, namely Adaptive Collision Free MAC (ACFM) and Cluster-Based RSU Centric Channel Access (CBRC) [44]. The combined approach of contention free and contention-based accesses can be divided into seven types, namely CSMA and Self-Organizing TDMA MAC (CS-TDMA), Space-Orthogonal Frequency-Time Medium Access Control (SOFT-MAC), Hybrid Efficient and Reliable MAC (HER-MAC), Dedicated Multichannel MAC with Adaptive Broadcasting (DMMAC), Clustering-Based Multichannel MAC (CBMMAC), Cluster-Based Medium Access Control Protocol (CBMCS), and Risk-Aware Dynamic MAC (R-MAC) [49].

Liu et al. [67] have proposed energy-efficient hybrid MAC protocols in a single window. In the case of hybrid MAC protocols, protocols based on CSMA/CA and TDMA access techniques provide better performance compared to CSMA/CA based MAC protocols. Hybrid protocols based on FDMA and CDMA better improve the network scalability than the pure FDMA and CDMA protocols. However, the disadvantages of FDMA-based and CDMA-based hybrid protocols are the requirement for expensive hardware and complicated operation and the requirement for power control. As a result, TDMA-based hybrid protocols are the most appealing in the context of IoT enabled communications [67].

5.5 Framework on CAVS

All the aforementioned MAC protocols are designed from a general communication point of view. However, MAC on CAVS should be designed to cater specific needs and requirements of vehicular applications. More specifically, the MAC design for a CAVS communication platform depends on the following factors:

Network Components The CAVS consists of various types of components such as onboard sensor nodes for intra-vehicular communications, roadside unit (RSU), which is installed along the road and provides real-time data services to passing vehicles, and different long-range communication enabled devices that can be

Table 5.2 Existing hybrid protocols

References	MAC name	Channels	Mobility	Density	Traffic characteristics	Multi-media application	Real-time application	Data traffic	Experiment	
									Analytical	Simulator
Dang et al.[31]	HERMAC	Multiple	NA	Low	Bidirectional	Yes	Yes	High load	NA	MATLAB
Torabi and Ghahfarokhi [110]	CSTDMA	Multiple	NA	Medium	Bidirectional	Yes	Yes	High load	Packet loss probability	MATLAB
Rezazade et al. [99]	STDMA	Single	High	High	Bidirectional	No	Yes	High load	NA	MATLAB
Wu et al. [119]	SOFTMAC	Single	NA	Low	Unidirectional	Yes	Yes	High load	NA	NA
Lu et al. [69]	DMMAC	Multiple	High	Medium	Unidirectional	NA	Yes	High load	NA	NS-2
Jayaraj et al. [49]	CBMMAC	Multiple	High	Low	Unidirectional	Yes	Yes	Medium	Probabilistic analysis	MATLAB
Nabi et al. [76]	CBMCS	Multiple	High	Low	Unidirectional	Yes	Yes	Medium	NA	C++
Guo et al. [42]	RMAC	Single	High	High	Unidirectional	No	Yes	High	Stochastic collision prevention model	NS-2
Nguyen et al. [78]	HTCMAC	Multiple	NA	Low	Bidirectional	NA	NA	Low, medium	Markov chain based Low,	MATLAB
Cao et al. [23]	Dynamic MAC	Single	NA	Low, high	Unidirectional	NA	NA	Low, medium, high	NA	MATLAB

supported by different technologies like WiMAX or 3G/4G for inter-vehicular communications [50, 92]. It shall be noted that cyber–physical interactions within the CAVS can be divided into two categories, namely intra-vehicular CPS and inter-vehicular CPS. The intra-vehicular CPS provides the kinetic performance of a single vehicle by combining and coordinating all of its components such as various types on board sensor nodes, actuators, and other smart devices into the complex environment. On the other hand, inter-vehicular communication incorporates V2V and V2I communications in which the traffic and vehicular networking are controlled from a CPS design standpoint.

Network Architecture The architecture of CAVS platform can be considered by reviewing some straightforward issues from the vehicular networking perspectives. Recall that there are two types of communications in such an environment, namely V2V [21] and V2I [25]. In V2V communications, vehicles communicate with one other using dedicated short-range communication standards such as DSRC/WAVE, radio frequency identifications (RFID), ZigBee, and Bluetooth [36, 83]. In V2I communications, the vehicles may use medium- or long-range communication technologies such as WiFi, WiMAX, or LTE/LTE-Advance for accessing the resources as shown in Fig. 5.4. The vehicles have specialized units such as onboard units (OBUs), application units, and sensors to communicate with one another and with the nearest fixed base access points. In the intra-vehicular communication, various short-range wireless communications such as Bluetooth, ZigBee, or UWB can be used [13, 68]. A typical network topology for intra-vehicular communication is of star shape. Alternatively, the mesh or cluster tree topology can also be

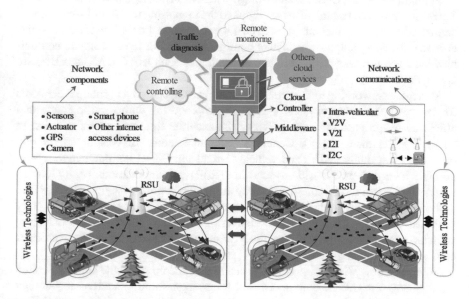

Fig. 5.4 A complete scenario of CAVS platform

employed at the expense of increasing complexity of the network architecture for further maintenance [83]. The long-range communications used for V2V, V2I, infrastructure-to-infrastructure (I2I), and infrastructure-to-controller (I2C) data exchange typically employ mesh topology for their communications. The RSU accesses all the vehicular information from OBU and passes it to the control center through a dedicated middleware that is responsible for maintaining the privacy and security.

CAVS is an integration of physical process and cyber systems via heterogeneous networking and communications. A typical CAVS is shown in Fig. 5.4, which comprises physical components (such as vehicles, DSRC/mobile devices, tablets), cyber systems (e.g., data center, traffic control center), and communications (e.g., vehicular networks).

Communication Technologies
In the CAVS for intelligent transportation systems, vehicles are expected to be able to compute and process the traffic information, and communicate that information with other vehicles, pedestrian, or roadside units (RSUs) using V2V or V2I communications in order to avoid traffic accidents and congestion [115]. For communications among vehicles or between vehicles and RSUs, wireless technology appears to be the most suitable option because the vehicular network topology changes quickly with the speed variation of the vehicles. In order to meet the communication requirements, each vehicle needs to dynamically change operating parameters needed for resilient communications based on the requirements and applications that are intended to support [19].

For intra-vehicular CPS, the short-range wireless technologies such as Bluetooth, ZigBee, or WiFi are suitable. However, due to the increasing number of sensor nodes demanded by intra-vehicular applications, a careful selection of such technologies is crucial for designing intra-vehicular communication. For inter-vehicular communication CPS, the long-range wireless technologies such as WiMAX and LTE are appropriate.

Vehicular networking and communications are considered as principal elements for the vehicular cyber-physical systems to improve the entire traffic safety and efficiency by propagating and analyzing the accurate time-critical information in an appropriate manner [94]. Generally, communications in vehicular CPS rely on vehicle-to-vehicle (V2V) and vehicle-to-roadside (V2R) communications with possible intermediate roadside-to-roadside (R2R) communications [92]. Conventional solutions to these issues use mainly automatic control systems using OBU in individual vehicles without any interaction with other vehicles. However, a recently proposed vehicular communication platform could help coordinate participating vehicles more efficiently and effectively with the assistance from inter-vehicular communications using V2V and/or V2R networking [12].

From the above discussion, it can be stated that according to the communication requirements for CAVS, both short- and long-range communications technology should be considered by addressing the following performance criteria, namely:

- Flexibility and Interoperability
- Safety
- Preparation for the management of quality of service (QoS)
- High Throughput
- Easy Installation and maintenance
- Mobility
- Low Cost
- Coverage

5.6   Open Research Issues

CAVS is still at its primary stage of development towards fully autonomous operations. Key development directions include proper improvement of onboard embedded systems, sensor networks, and communication systems. To develop a fully functioning CAVS platform, some key challenges in the following need to be addressed:

Privacy and Cyber-Security Privacy and security are major problems faced by vehicular communications. Whenever vehicles identification is utilized, the system can be safeguarded by making the involved parties accountable. Nevertheless, whenever vehicle authentication is used, the privacy of owner or driver/renter may be comprised. This basically means that, in vehicular systems, it is important to confirm the identity of the authenticating vehicle to maintain integrity regarding the supplied information. Cyber-attacks are the main security concern in the wireless network architecture used for CAVS. The security vulnerabilities of the vehicular network can cause damages and fatalities to the vehicle and driver.

Heterogeneous Wireless Connectivity The CAVS is integrated with various vehicular devices or access technologies. Heterogeneous wireless connectivity comprises a large number of specialized sensor nodes, in which a subset of the nodes can dynamically set up a self-organizing communication network. Effective management and integration of different heterogeneous networks constitutes to one of the primary challenges in improving the overall quality of service (QoS) in networks with different short-range or long-range wireless connectivity that can access trust information from the vehicular platform.

Delay Sensitivity and QoS The CAVS wireless connectivity is considered one of the most important enabling factors to deliver time-critical communications in a very short time frame.

Platform Independence and Interoperability The CAVS is supposed to be fully automatic and interoperable across different platforms with complex tasks and environments. It is challenging to achieve this where it is required from the manufacturing point of view to assemble different components and systems from different vendors and suppliers.

Small Delay and High-Speed Communication Technology Communication technology for CAVS should have a very small delay and latency requirement (i.e., in the order of microsecond or less) so that sensing and processed information can be used to stabilize the system in a timely manner. This requirement has not been fully met by the existing CAVS-wireless technologies.

High Data Rates The limited data rate of existing wireless technologies is insufficient to manage the requirement of CAVS because of high computational scenarios like videos. It is therefore important to improve existing data rates across all the CAVS-wireless access technologies.

5.7 Conclusion

In this chapter we have studied the CAVS from a MAC layer perspective. We have discussed the incorporation of the concept of IoT on vehicular systems and its impact based on statistical information. In order to address the MAC layer's challenges and issues on CAVS, the detailed state of art was discussed. A framework for CAVS was illustrated and finally the open research issues were described.

Acknowledgement This work is partially supported by the project RDU150391 funded by University Malaysia Pahang.

References

1. A.M. Abbas, J. Ali, M.A. Rahman, S. Azad, Paper presented at the 2016 international conference on comparative investigation on CSMA/CA-based MAC protocols for scalable networks computer and communication engineering (ICCCE) (2016)
2. H. Abid, L.T.T. Phuong, J. Wang, S. Lee, S. Qaisar, V-Cloud: vehicular cyber-physical systems and cloud computing. Paper presented at the proceedings of the 4th international symposium on applied sciences in biomedical and communication technologies (2011)
3. M. Ahmed, C.U. Saraydar, T. ElBatt, J. Yin, T. Talty, & M. Ames, Intra-vehicular wireless networks. Paper presented at the Globecom workshops, IEEE, New York (2007)
4. A. Aijaz, A.H. Aghvami, Cognitive machine-to-machine communications for internet-of-things: a protocol stack perspective. IEEE Internet Things J. **2**(2), 103–112 (2015)
5. A. Al-Fuqaha, M. Guizani, M. Mohammadi, M. Aledhari, M. Ayyash, Internet of things: a survey on enabling technologies, protocols, and applications. IEEE Commun Surveys Tutorials **17**(4), 2347–2376 (2015)
6. T. AlSkaif, B. Bellalta, M.G. Zapata, J.M.B. Ordinas, Energy efficiency of MAC protocols in low data rate wireless multimedia sensor networks: a comparative study. Ad Hoc Netw **56**, 141–157 (2016)
7. A.S. Althobaiti, M. Abdullah, Medium access control protocols for wireless sensor networks classifications and cross-layering. Proc. Comp. Sci. **65**, 4–16 (2015)
8. T. Anand, K. Banupriya, M. Deebika, A. Anusiya, T. Anand, K. Banupriya, et al., Intelligent transportation systems using iot service for vehicular data cloud. Int J Innov Res Sci Technol **2**(2), 80–86 (2015)

9. M. Arifuzzaman, M. Matsumoto, T. Sato, An intelligent hybrid MAC with traffic-differentiation-based QoS for wireless sensor networks. IEEE Sensors J. **13**(6), 2391–2399 (2013)
10. M.S. Azad, M.M. Uddin, F. Anwar, M.A. Rahman, Performance evaluation of wireless routing protocols in mobile wimax environment. Paper presented at the proceedings of the international multiconference of engineers and computer scientists (2008)
11. A. Bachir, M. Dohler, T. Watteyne, K.K. Leung, MAC essentials for wireless sensor networks. IEEE Commun Surv Tutorial **12**(2), 222–248 (2010)
12. R.S. Bali, N. Kumar, Secure clustering for efficient data dissemination in vehicular cyber–physical systems. Futur. Gener. Comput. Syst. **56**, 476–492 (2016)
13. C.U. Bas, S.C. Ergen, Ultra-wideband channel model for intra-vehicular wireless sensor networks beneath the chassis: from statistical model to simulations. IEEE Trans. Veh. Technol. **62**(1), 14–25 (2013)
14. J. Beaudaux, A. Gallais, J. Montavont, T. Noel, D. Roth, E. Valentin, Thorough empirical analysis of X-MAC over a large scale internet of things testbed. IEEE Sensors J. **14**(2), 383–392 (2014)
15. O. Bello, S. Zeadally, Intelligent device-to-device communication in the internet of things. IEEE Syst. J. **10**(3), 1172–1182 (2016)
16. O. Bello, S. Zeadally, M. Badra, Network layer inter-operation of device-to-device communication technologies in internet of things (IoT). Ad Hoc Netw. **57**, 52–62 (2017)
17. M.Z.A. Bhuiyan, G. Wang, J. Cao, J. Wu, Deploying wireless sensor networks with fault-tolerance for structural health monitoring. IEEE Trans. Comput. **64**(2), 382–395 (2015)
18. M.Z.A. Bhuiyan, G. Wang, J. Wu, J. Cao, X. Liu, T. Wang, Dependable structural health monitoring using wireless sensor networks. IEEE Trans Depend Secure Comput **14**(4), 363–376 (2017)
19. S. Bitam, A. Mellouk, S. Zeadally, VANET-cloud: a generic cloud computing model for vehicular ad hoc networks. IEEE Wirel. Commun. **22**(1), 96–102 (2015)
20. T.M. Bojan, U.R. Kumar, V.M. Bojan, An internet of things based intelligent transportation system. Paper presented at the 2014 IEEE international conference on vehicular electronics and safety (ICVES) (2014)
21. A. Burg, A. Chattopadhyay, K.-Y. Lam, Wireless communication and security issues for cyber–physical systems and the internet-of-things. Proc. IEEE **106**(1), 38–60 (2018)
22. A.S. Cacciapuoti, M. Caleffi, L. Paura, M.A. Rahman, Link quality estimators for multi-hop mesh network. Paper presented at the Euro Med Telco conference (EMTC) (2014)
23. L. Cao, W. Xu, X. Lin, J. Lin, A CSMA/TDMA dynamic splitting scheme for MAC protocol in VANETs. Paper presented at the 2013 international conference on Wireless Communications & Signal Processing (WCSP) (2013)
24. R.C. Carrano, D. Passos, L.C. Magalhaes, C.V. Albuquerque, Survey and taxonomy of duty cycling mechanisms in wireless sensor networks. IEEE Commun Surv Tutorial **16**(1), 181–194 (2014)
25. D. Centea, I. Singh, M. Elbestawi, Framework for the development of a cyber-physical systems learning centre, in *Online Engineering & Internet of Things*, (Springer, Cham, 2018), pp. 919–930
26. D. Cerotti, S. Distefano, G. Merlino, A. Puliafito, A crowd-cooperative approach for intelligent transportation systems. IEEE Trans. Intell. Transp. Syst. **18**(6), 1529–1539 (2017)
27. K. Chen, M. Ma, E. Cheng, F. Yuan, W. Su, A survey on MAC protocols for underwater wireless sensor networks. IEEE Commun Surv Tutorial **16**(3), 1433–1447 (2014)
28. Q. Chi, H. Yan, C. Zhang, Z. Pang, L. Da Xu, A reconfigurable smart sensor interface for industrial WSN in IoT environment. IEEE Trans Industr Inform **10**(2), 1417–1425 (2014)
29. S. Climent, A. Sanchez, J.V. Capella, N. Meratnia, J.J. Serrano, Underwater acoustic wireless sensor networks: Advances and future trends in physical, MAC and routing layers. Sensors **14**(1), 795–833 (2014)
30. J. Contreras-Castillo, S. Zeadally, J.A. Guerrero Ibáñez, A seven-layered model architecture for internet of vehicles. J Inform Telecommun **1**(1), 4–22 (2017)

31. D.N.M. Dang, H.N. Dang, V. Nguyen, Z. Htike, C.S. Hong. HER-MAC: A hybrid efficient and reliable MAC for vehicular ad hoc networks. Paper presented at the 2014 IEEE 28th international conference on advanced information networking and applications (AINA) (2014)

32. F.Z. Djiroun, D. Djenouri, MAC protocols with wake-up Radio for Wireless Sensor Networks: a review. IEEE Commun Surv Tutorial **19**(1), 587–618 (2017)

33. M. Dong, K. Ota, A. Liu, RMER: reliable and energy-efficient data collection for large-scale wireless sensor networks. IEEE Internet Things J. **3**(4), 511–519 (2016)

34. M. Doudou, D. Djenouri, N. Badache, A. Bouabdallah, Synchronous contention-based MAC protocols for delay-sensitive wireless sensor networks: a review and taxonomy. J. Netw. Comput. Appl. **38**, 172–184 (2014)

35. T. ElBatt, C. Saraydar, M. Ames, T. Talty, Potential for intra-vehicle wireless automotive sensor networks. Paper presented at the 2006 IEEE Sarnoff symposium (2006)

36. Y. Feng, B. Hu, H. Hao, Y. Gao, Z. Li, J. Tan, Design of distributed cyber-physical systems for connected and automated vehicles with implementing methodologies. IEEE Trans Indus Inform **14**(9), P4200–P4211 (2018)

37. M.H.S. Gilani, I. Sarrafi, M. Abbaspour, An adaptive CSMA/TDMA hybrid MAC for energy and throughput improvement of wireless sensor networks. Ad Hoc Netw. **11**(4), 1297–1304 (2013)

38. S. Gill, P. Sahni, P. Chawla, S. Kaur, Intelligent transportation architecture for enhanced security and integrity in vehicles integrated internet of things. Indian J. Sci. Technol. **10**(10), 1–5 (2017)

39. S.A. Gopalan, D.-H. Kim, J.-W. Nah, J.-T. Park, A survey on power-efficient MAC protocols for wireless body area networks. Paper presented at the 3rd IEEE international conference on broadband network and multimedia technology (IC-BNMT) (2010)

40. S.A. Gopalan, J.-T. Park, Energy-efficient MAC protocols for wireless body area networks: survey. Paper presented at the 2010 international congress on ultra modern telecommunications and control systems and workshops (ICUMT) (2010)

41. J.A. Guerrero-ibanez, S. Zeadally, J. Contreras-Castillo, Integration challenges of intelligent transportation systems with connected vehicle, cloud computing, and internet of things technologies. IEEE Wirel. Commun. **22**(6), 122–128 (2015)

42. W. Guo, L. Huang, L. Chen, H. Xu, C. Miao, R-mac: risk-aware dynamic mac protocol for vehicular cooperative collision avoidance system. Int J Distributed Sensor Netw **9**(5), 686713 (2013)

43. N. Gupta, A. Prakash, R. Tripathi, Medium access control protocols for safety applications in vehicular ad-hoc network: a classification and comprehensive survey. Veh Commun **2**(4), 223–237 (2015)

44. M. Hadded, P. Muhlethaler, A. Laouiti, R. Zagrouba, L.A. Saidane, TDMA-based MAC protocols for vehicular ad hoc networks: a survey, qualitative analysis, and open research issues. IEEE Commun Surv Tutorial **17**(4), 2461–2492 (2015)

45. W. He, G. Yan, L. Da Xu, Developing vehicular data cloud services in the IoT environment. IEEE Trans Indus Inform **10**(2), 1587–1595 (2014)

46. V.J. Hodge, S. O'Keefe, M. Weeks, A. Moulds, Wireless sensor networks for condition monitoring in the railway industry: a survey. IEEE Trans. Intell. Transp. Syst. **16**(3), 1088–1106 (2015)

47. P. Huang, L. Xiao, S. Soltani, M.W. Mutka, N. Xi, The evolution of MAC protocols in wireless sensor networks: a survey. IEEE Commun Surv Tutorial **15**(1), 101–120 (2013)

48. P. Hurni, T. Braun, Maxmac: a maximally traffic-adaptive mac protocol for wireless sensor networks. Paper presented at the European conference on wireless sensor networks (2010)

49. V. Jayaraj, C. Hemanth, R. Sangeetha, A survey on hybrid MAC protocols for vehicular ad-hoc networks. Veh Commun **6**, 29–36 (2016)

50. D. Jia, K. Lu, J. Wang, On the network connectivity of platoon-based vehicular cyber-physical systems. Transport Res Part C: EmerTechnol **40**, 215–230 (2014)

51. D. Jia, K. Lu, J. Wang, X. Zhang, X. Shen, A survey on platoon-based vehicular cyber-physical systems. IEEE Commun Surv Tutorial **18**(1), 263–284 (2016)
52. Y. Jiang, S. Yin, Recursive total principle component regression based fault detection and its application to vehicular cyber-physical systems. IEEE Trans Indus Inform **14**(4), 1415–1423 (2017)
53. J. Jin, J. Gubbi, S. Marusic, M. Palaniswami, An information framework for creating a smart city through internet of things. IEEE Internet Things J. **1**(2), 112–121 (2014)
54. A. Kakria, T.C. Aseri, Survey of synchronous MAC protocols for wireless sensor networks. Paper presented at the 2014 recent advances in engineering and computational sciences (RAECS) (2014)
55. P. Kamalinejad, C. Mahapatra, Z. Sheng, S. Mirabbasi, V.C. Leung, Y.L. Guan, Wireless energy harvesting for the internet of things. IEEE Commun. Mag. **53**(6), 102–108 (2015)
56. S.D.T. Kelly, N.K. Suryadevara, S.C. Mukhopadhyay, Towards the implementation of IoT for environmental condition monitoring in homes. IEEE Sensors J. **13**(10), 3846–3853 (2013)
57. A.A. Khan, M.H. Rehmani, M. Reisslein, Cognitive radio for smart grids: survey of architectures, spectrum sensing mechanisms, and networking protocols. IEEE Commun Surv Tutorial **18**(1), 860–898 (2016)
58. R.A.M. Khan, H. Karl, MAC protocols for cooperative diversity in wireless LANs and wireless sensor networks. IEEE Commun Surv Tutorial **16**(1), 46–63 (2014)
59. J. Kim, J. Lee, J. Kim, J. Yun, M2M service platforms: Survey, issues, and enabling technologies. IEEE Commun Surv Tutorial **16**(1), 61–76 (2014)
60. N. Kumar, M. Singh, S. Zeadally, J.J. Rodrigues, S. Rho, Cloud-assisted context-aware vehicular cyber-physical system for PHEVs in smart grid. IEEE Syst. J. **11**(1), 140–151 (2017)
61. M. Kumaraswamy, K. Shaila, V. Tejaswi, K. Venugopal, S. Iyengar, L. Patnaik, QoS driven distributed multi-channel scheduling MAC protocol for multihop WSNs. Paper presented at the 2014 international conference on computer and communication technology (ICCCT) (2014)
62. I. Lee, K. Lee, The internet of things (IoT): applications, investments, and challenges for enterprises. Bus. Horiz. **58**(4), 431–440 (2015)
63. X. Li, X. Yu, A. Wagh, C. Qiao, Human factors-aware service scheduling in vehicular cyber-physical systems. Paper presented at the 2011 proceedings IEEE INFOCOM (2011)
64. R. Liao, B. Bellalta, M. Oliver, Z. Niu, MU-MIMO MAC protocols for wireless local area networks: a survey. IEEE Commun Surv Tutorial **18**(1), 162–183 (2016)
65. J.-R. Lin, T. Talty, O.K. Tonguz, A blind zone alert system based on intra-vehicular wireless sensor networks. IEEE Trans Indus Inform **11**(2), 476–484 (2015a)
66. J.-R. Lin, T. Talty, O.K. Tonguz, On the potential of bluetooth low energy technology for vehicular applications. IEEE Commun. Mag. **53**(1), 267–275 (2015b)
67. Y. Liu, C. Yuen, X. Cao, N.U. Hassan, J. Chen, Design of a scalable hybrid MAC protocol for heterogeneous M2M networks. IEEE Internet Things J. **1**(1), 99–111 (2014)
68. N. Lu, N. Cheng, N. Zhang, X. Shen, J.W. Mark, Connected vehicles: solutions and challenges. IEEE Internet Things J. **1**(4), 289–299 (2014)
69. N. Lu, Y. Ji, F. Liu, X. Wang, A dedicated multi-channel MAC protocol design for VANET with adaptive broadcasting. Paper presented at the 2010 IEEE wireless communications and networking conference (WCNC) (2010)
70. L. Mainetti, L. Patrono, A. Vilei, Evolution of wireless sensor networks towards the internet of things: a survey. Paper presented at the 2011 19th international conference on software, telecommunications and computer networks (SoftCOM) (2011)
71. T. Maitra, S. Roy, A comparative study on popular MAC protocols for mixed wireless sensor networks: from implementation viewpoint. Comput Sci Rev **22**, 107–134 (2016)
72. R.T. Matani, T.M. Vasavada, A survey on MAC protocols for data collection in wireless sensor networks. Int J Comput Appl **114**(6), 4–7 (2015)
73. A. Miloslavov, M. Veeraraghavan, Sensor data fusion algorithms for vehicular cyber-physical systems. IEEE Trans Parallel Distribut Syst **23**(9), 1762–1774 (2012)

74. S. Mishra, R.R. Swain, T.K. Samal, M.R. Kabat, CS-ATMA: a hybrid single channel MAC layer protocol for wireless sensor networks, in *Computational Intelligence in Data Mining*, vol. 3, (Springer, New York, 2015), pp. 271–279

75. P.D. Mitchell, J. Qiu, H. Li, D. Grace, Use of aerial platforms for energy efficient medium access control in wireless sensor networks. Comput. Commun. **33**(4), 500–512 (2010)

76. M. Nabi, M. Geilen, T. Basten, M. Blagojevic, Efficient cluster mobility support for TDMA-based MAC protocols in wireless sensor networks. ACM Trans Sensor Netw (TOSN) **10**(4), 65 (2014)

77. A.H. Ngu, M. Gutierrez, V. Metsis, S. Nepal, Q.Z. Sheng, IoT middleware: a survey on issues and enabling technologies. IEEE Internet Things J. **4**(1), 1–20 (2017)

78. V. Nguyen, T.Z. Oo, P. Chuan, C.S. Hong, An efficient time slot acquisition on the hybrid TDMA/CSMA multichannel MAC in VANETs. IEEE Commun. Lett. **20**(5), 970–973 (2016)

79. K. Ovsthus, L.M. Kristensen, An industrial perspective on wireless sensor networks—a survey of requirements, protocols, and challenges. IEEE Commun Surv Tutorial **16**(3), 1391–1412 (2014)

80. J. Park, C. Lee, J.-H. Park, B-c. Choi, & J. G Ko,. Poster: exploiting wireless CAN bus bridges for intra-vehicle communications. Paper presented at the 2014 IEEE vehicular networking conference (VNC) (2014)

81. G. Parsons, Standardizing machine-to-machine (M2M) communications. IEEE Commun. Mag. **54**(12), 2–3 (2016)

82. A. Rahim, N. Javaid, M. Aslam, Z. Rahman, U. Qasim, Z. Khan, A comprehensive survey of MAC protocols for wireless body area networks. Paper presented at the 2012 seventh international conference on broadband, wireless computing, communication and applications (BWCCA) (2012)

83. M. Rahman, J. Ali, M.N. Kabir, S. Azad, A performance investigation on IoT enabled intra-vehicular wireless sensor networks. Int J Autom Mech Eng **14**(1), 3970–3984 (2017)

84. M.A. Rahman, Design of wireless sensor network for intra-vehicular communications. Paper presented at the international conference on wired/wireless internet communications (2014a)

85. M.A. Rahman, Reliability analysis of ZigBee based intra-vehicle wireless sensor networks. Paper presented at the international workshop on communication technologies for vehicles (2014b)

86. M.A. Rahman, M.S. Azad, F. Anwar, Intergating multiple metrics to improve the performance of a routing protocol over wireless mesh networks. Paper presented at the 2009 international conference on signal processing systems (2009)

87. M.A. Rahman, M.N. Kabir, S. Azad, J. Ali, On mitigating hop-to-hop congestion problem in IoT enabled intra-vehicular communication. Paper presented at the 2015 4th international conference on software engineering and computer systems (ICSECS) (2015)

88. A. Rai, S. Deswal, P. Singh, MAC protocols in wireless sensor network: A survey. Int J New Innov Eng Technol **5**(1), 95–101 (2016)

89. A. Rajandekar, B. Sikdar, A survey of MAC layer issues and protocols for machine-to-machine communications. IEEE Internet Things J. **2**(2), 175–186 (2015)

90. Y. Rao, Y.-m. Cao, C. Deng, Z.-h. Jiang, J. Zhu, L.-y. Fu, R.-c. Wang, Performance analysis and simulation verification of S-MAC for wireless sensor networks. Comput Electr Eng **56**, 468–484 (2016)

91. M.B. Rasheed, N. Javaid, M. Imran, Z.A. Khan, U. Qasim, A. Vasilakos, Delay and energy consumption analysis of priority guaranteed MAC protocol for wireless body area networks. Wirel. Netw **23**(4), 1249–1266 (2017)

92. D.B. Rawat, C. Bajracharya, Adaptive connectivity for vehicular cyber-physical systems, in *Vehicular Cyber Physical Systems*, (Springer, Cham, 2017a), pp. 15–24

93. D.B. Rawat, C. Bajracharya, An overview of vehicular networking and cyber-physical systems, in *Vehicular Cyber Physical Systems*, (Springer, Cham, 2017b), pp. 1–13

94. D.B. Rawat, C. Bajracharya, *Vehicular Cyber Physical Systems* (Springer, Cham, 2017c)

95. S. Ray, I. Demirkol, W. Heinzelman, Supporting bursty traffic in wireless sensor networks through a distributed advertisement-based TDMA protocol (ATMA). Ad Hoc Netw. **11**(3), 959–974 (2013)
96. A. Razaque, K.M. Elleithy, Energy-efficient boarder node medium access control protocol for wireless sensor networks. Sensors **14**(3), 5074–5117 (2014)
97. A. Razaque, K.M. Elleithy, Low duty cycle, energy-efficient and mobility-based boarder node—MAC hybrid protocol for wireless sensor networks. J Sig Process Syst **81**(2), 265–284 (2015)
98. S. Reis, D. Pesch, B.-L. Wenning, M. Kuhn, Intra-vehicle wireless sensor network communication quality assessment via packet delivery ratio measurements. Paper presented at the international conference on mobile networks and management (2016)
99. L. Rezazade, H.S. Aghdasi, S.A. Ghorashi, M. Abbaspour, A novel STDMA MAC protocol for vehicular ad-hoc networks. Paper presented at the 2011 international symposium on computer networks and distributed systems (CNDS) (2011)
100. M. Salajegheh, H. Soroush, A. Kalis, HYMAC: Hybrid TDMA/FDMA medium access control protocol for wireless sensor networks. Paper presented at the IEEE 18th international symposium on personal, indoor and mobile radio communications. PIMRC 2007 (2007)
101. Z. Sheng, S. Yang, Y. Yu, A. Vasilakos, J. Mccann, K. Leung, A survey on the ietf protocol suite for the internet of things: Standards, challenges, and opportunities. IEEE Wirel. Commun. **20**(6), 91–98 (2013)
102. L. Shi, A.O. Fapojuwo, TDMA scheduling with optimized energy efficiency and minimum delay in clustered wireless sensor networks. IEEE Trans. Mob. Comput. **9**(7), 927–940 (2010)
103. B. Shrestha, E. Hossain, K.W. Choi, Distributed and centralized hybrid CSMA/CA-TDMA schemes for single-hop wireless networks. IEEE Trans. Wirel. Commun. **13**(7), 4050–4065 (2014)
104. M. Shu, D. Yuan, C. Zhang, Y. Wang, C. Chen, A MAC protocol for medical monitoring applications of wireless body area networks. Sensors **15**(6), 12906–12931 (2015)
105. J. Song, A. Kunz, R.R.V. Prasad, Z. Sheng, R. Yu, Research to standards: next generation IoT/M2M applications, networks and architectures. IEEE Commun. Mag. **54**(12), 14–15 (2016)
106. B.L.R. Stojkoska, K.V. Trivodaliev, A review of internet of things for smart home: challenges and solutions. J. Clean. Prod. **140**, 1454–1464 (2017)
107. W. Sun, J. Liu, H. Zhang, When smart wearables meet intelligent vehicles: challenges and future directions. IEEE Wirel. Commun. **24**(3), 58–65 (2017)
108. P. Suriyachai, U. Roedig, A. Scott, A survey of MAC protocols for mission-critical applications in wireless sensor networks. IEEE Commun Surv Tutorial **14**(2), 240–264 (2012)
109. R.R. Swain, S. Mishra, T.K. Samal, M.R. Kabat, An energy efficient advertisement based multichannel distributed MAC protocol for wireless sensor networks (Adv-MMAC). Wirel. Pers. Commun. **95**(2), 655–682 (2017)
110. N. Torabi, B.S. Ghahfarokhi, A bandwidth-efficient and fair CSMA/TDMA based multichannel MAC scheme for V2V communications. Telecommun. Syst. **64**(2), 367–390 (2017)
111. S. Tuohy, M. Glavin, C. Hughes, E. Jones, M. Trivedi, L. Kilmartin, Intra-vehicle networks: A review. IEEE Trans. Intell. Transp. Syst. **16**(2), 534–545 (2015)
112. P. Tuset-Peiro, F. Vazquez-Gallego, J. Alonso-Zarate, L. Alonso, X. Vilajosana, LPDQ: A self-scheduled TDMA MAC protocol for one-hop dynamic low-power wireless networks. Perv Mobile Comput **20**, 84–99 (2015)
113. L. Uden, L. Uden, W. He, W. He, How the internet of things can help knowledge management: a case study from the automotive domain. J. Knowl. Manag. **21**(1), 57–70 (2017)
114. P. Vlacheas, R. Giaffreda, V. Stavroulaki, D. Kelaidonis, V. Foteinos, G. Poulios, et al., Enabling smart cities through a cognitive management framework for the internet of things. IEEE Commun. Mag. **51**(6), 102–111 (2013)
115. J. Wan, D. Zhang, S. Zhao, L. Yang, J. Lloret, Context-aware vehicular cyber-physical systems with cloud support: Architecture, challenges, and solutions. IEEE Commun. Mag. **52**(8), 106–113 (2014)

116. C. Wang, Z. Zhao, L. Zhu, H. Yao, An energy efficient routing protocol for in-vehicle wireless sensor networks. Paper presented at the international conference of pioneering computer scientists, engineers and educators (2017)
117. S. Woo, H.J. Jo, I.S. Kim, D.H. Lee, A practical security architecture for in-vehicle CAN-FD. IEEE Trans. Intell. Transp. Syst. **17**(8), 2248–2261 (2016)
118. S. Woo, H.J. Jo, D.H. Lee, A practical wireless attack on the connected car and security protocol for in-vehicle CAN. IEEE Trans. Intell. Transp. Syst. **16**(2), 993–1006 (2015)
119. H. Wu, Y. Liu, Q. Zhang, Z.-L. Zhang, SoftMAC: Layer 2.5 collaborative MAC for multimedia support in multihop wireless networks. IEEE Trans Mob Comput **6**(1), 12–25 (2007)
120. B. Yahya, J. Ben-Othman, Towards a classification of energy aware MAC protocols for wireless sensor networks. Wirel. Commun. Mob. Comput. **9**(12), 1572–1607 (2009)
121. M. Yigit, O. Durmaz Incel, S. Baktir, V.C. Gungor, QoS-aware MAC protocols utilizing sectored antenna for wireless sensor networks-based smart grid applications. Int. J. Commun. Syst. **30**(7), e3168 (2017)
122. C. Yin, Y. Li, D. Zhang, Y. Cheng, M. Yin, DSMAC: An energy-efficient MAC protocol in event-driven sensor networks. Paper presented at the 2nd international conference on advanced computer control (ICACC) (2010)
123. Y. Zhang, B. Chen, X. Lu, Intelligent monitoring system on refrigerator trucks based on the internet of things. Paper presented at the international conference on wireless communications and applications (2011)
124. Y. Zhao, M. Ma, C. Miao, T. Nguyen, An energy-efficient and low-latency MAC protocol with adaptive scheduling for multi-hop wireless sensor networks. Comput. Commun. **33**(12), 1452–1461 (2010)
125. H. Zhou, B. Liu, D. Wang, Design and research of urban intelligent transportation system based on the internet of things. Internet of Things **312**, 572–580 (2012)

Chapter 6
A CPS-Enhanced Subway Operations Safety System Based on the Short-Term Prediction of the Passenger Flow

Shaobo Zhong, Zhi Xiong, Guannan Yao, and Wei Zhu

6.1 Introduction

The subway transport is an important part of the transportation system that helps limit the challenges created by rapid urbanization and traffic congestion [1]. Urban subways around the world have undergone significant development, which has created novel challenges to subway operations safety management and emergency response systems. For example, Beijing built its first subway line in 1971. By the end of 2016, a total of 19 subway lines had been built and put into operation, and these lines manage over 50% of the whole city's passenger flow. Nevertheless, the public transportation capacity of Beijing's subways is still far behind the development of the rest of the city. Subway scheduling and safety issues have become serious problems. In a subway system, transfer stations are the focus of subway operations and staff due to the great number of inbound and outbound passengers, as well as passengers transferring to other lines. The accurate prediction of the passenger flow of subway stations (especially transfer stations) is important for real-time scheduling and the prevention of, and response to, emergencies.

Forecast types of subway passenger flow can generally be divided into long-term, medium-term, and short-term. For long-term predictions, the most commonly used approaches include the 4-step transportation planning model or regression techniques [2, 3], which estimate future flow patterns and travel demands based on contributing factors such as the economy, demography, transit characteristics, and land use type [4–11]. Statistical and computational intelligence (CI)-based models are popular for short-term and medium-term passenger flow prediction [12].

S. Zhong · W. Zhu (✉)
Beijing Research Center of Urban Systems Engineering, Beijing, P.R. China

Z. Xiong · G. Yao
Department of Engineering Physics, Tsinghua University, Beijing, P.R. China

© Springer Nature Switzerland AG 2020
S. Hu, B. Yu (eds.), *Big Data Analytics for Cyber-Physical Systems*,
https://doi.org/10.1007/978-3-030-43494-6_6

These models include the time series [13–15], neural network [16–18], support vector machine [19, 20], and empirical mode decomposition [21]. Classic methods for time series analysis analyze historical observations by identifying linear and nonlinear models and forecast values for the next time steps [22]. Passenger flows are inextricably bound up with time, and they are expressed as temporal sequences; thus, they can be modeled and predicted with time series models. The CI-based algorithms (e.g., support vector machine and neural network), which are the majority of the currently popular short-term ridership forecasting approaches, exhibit a great capability for analyzing highly nonlinear and complex phenomena with less rigorous assumptions and prerequisites than statistical models and often yield more accurate predictions [12]. In recent years, deep neural network models have become useful in a range of fields. Some researchers investigated the application of a deep neural network (DNN) in the short-term prediction of the passenger flow [23, 24]. DNNs generally yield better results than other types of tools because these models (e.g., the recurrent neural network (RNN) and the derived long short-term memory (LSTM) neural network) can fit complex nonlinear structures and depict the delays in time series. However, the explanatory power of CI-based approaches has been criticized for its weak interpretation and inference capabilities [1, 25]. Nonetheless, statistical approaches have advantages over the CI-based models in terms of brevity and efficiency. Even in some suitable conditions, statistical methods manifest better outcomes. Furthermore, they are sometimes preferred in particular situations (e.g., in emergency responses, where time is generally more important than other factors). Therefore, in practice, implementing various methods and selecting a condition-specific one should produce optimal predictions.

Regardless of the model applied, historical observations are typically used to train the models and output model parameters under given convergence conditions. Then, to predict the passenger flow, input parameter data need to be collected and input to the trained models. Under a short-term prediction scenario of the subway passenger flow, this is a real-time task that generally requires the aggregation and transmission of multi-source monitoring and forecasting variables. Thanks to the development of the Internet of Things (IoT) and cyber-physical systems (CPSs), heterogeneous sensor data can now be integrated into passenger traffic systems, processed, and applied. The prediction results and control signals can subsequently be forwarded to stakeholders or intelligent devices (e.g., ticket gates). Thus, an enhanced passenger traffic system based on the CPS can be built to deliver the intelligent safety operations management of subways.

The remainder of this paper is organized as follows. Section 6.2 reviews several popular models for the short-term prediction of the subway passenger flow. Section 6.3 introduces our CPS-enhanced safety management framework and the SAFETY (system, adjust, facilities, early warning, time control, and yielding) implementation. Section 6.4 discusses the test case of the Beijing subway. Section 6.5 concludes with a summary.

6.2   Short-Term Prediction Models of the Subway Passenger Flow

6.2.1   Time Series Forecast Methods

A time series is a series of observations obtained in a certain chronological order. There is a certain relationship between the adjacent values of the time series, and they depend on each other. Therefore, a time series also provides various rules for the prediction analysis. According to the different processing methods of historical data, time series analysis can be divided into two types: simple average and weighted average. In addition, time series predictions often need to simultaneously consider trends, periodicity, randomness, and synthesis. The autoregressive integrated moving average (ARIMA) model and exponential smoothing method are two of the most widely used techniques in time series analysis in the short-term prediction of the subway passenger flow.

(1) ARIMA. The ARIMA model is shown in Eq. (6.1) [26]:

$$\left(1 - \sum_{i=1}^{p} \phi_i L^i\right)(1 - L)^d y_t = \left(1 + \sum_{i=1}^{q} \theta_i L^i\right)\varepsilon_t \tag{6.1}$$

where y_t is the observation at time t; L is the lag operator; p and q are the autoregressive and moving average orders, respectively; d is the number of differences (order) made to make it a stationary sequence, $d \in Z, d > 0$; ϕ and θ are the undetermined coefficients; and ε_t is an independent random error term. ARIMA (p, d, q) is short for the ARIMA model.

(2) Exponential Smoothing. The exponential smoothing method belongs to the moving average method by giving different weights to past observations; that is, the closer observations are greater than the farther observations. The common feature of these methods is that the predicted value is the weighted sum of the previous observations, and different weights are assigned to different data. The weight of the new data is larger, and the weight of the old data is smaller [27].

The formula for exponential smoothing is shown in Eq. (6.2):

$$S_t = a y_t + (1 - a) S_{t-1} \tag{6.2}$$

where y_t, S_t is the observation and the smoothing statistic at time t, respectively. a is the smoothing factor.

Generally, time series prediction includes the following steps:

Step 1:   Identify series stationarity. If it is a non-stationary sequence, one should make it stationary first.

Step 2:   Select a suitable model and fit the model parameters according to the historical observations.

Step 3:   Verify and improve the model with verification data.
Step 4:   Apply the model for prediction.

6.2.2   Prediction Methods Incorporating Multiple Restriction Factors

While classic methods for time series analysis can directly estimate the model parameters from historical observations, they are difficult to use to model the relationship between the variables of interest and multiple restriction factors. For example, transportation policy, weather, and holidays have complex impacts on the passenger flow. Some CI-based methods were applied to the short-term prediction of the subway passenger flow. Among them, the artificial neural network techniques have gained extensive utilization since they can literally incorporate hierarchical indicators and model the nonlinear relationship between input and output. The BP neural network and Elman neural network are two of the classic CI-based methods.

(1) BP neural network. The back propagation (BP) neural network is characterized by its ability to adjust the error by continuously passing information forward. The basic BP algorithm includes two phases: signal forward propagation and error back propagation [28, 29]. The topological structure of the BP neural network is shown in Fig. 6.1.

   The model for the prediction of the subway passenger flow (daily) is based on the BP neural network and consists of an input layer, a hidden layer, and an output layer. The input layer includes the month, day of the week, weekend flag, weather state, highest daily temperature, lowest daily temperature, and important holiday flag. Of these seven nodes, the output layer has one node representing the daily passenger flow, and the number of hidden layer nodes can be determined with a variety of methods [30]. For example, the number of hidden layer nodes with the smallest training error can be selected from multiple trainings.

(2) Elman neural network. The Elman neural network is a feedback network with an additional transit operator. Unlike the BP neural network, the additional operator can play the roles of both memory and delay [31, 32]. As shown in Fig. 6.2, the Elman network has **tansig** neurons in its hidden (recurrent) layer and **purelin** neurons in its output layer. The former is a nonlinear activation function, and the latter is a linear activation function.

The model for the prediction of the subway daily passenger flow, which is based on the Elman neural network, is similar to the BP neural network. The same input nodes and output nodes were adopted in the modeling, training, and predicting.

Table 6.1 presents the seven input nodes used in the BP neural network and Elman neural network.

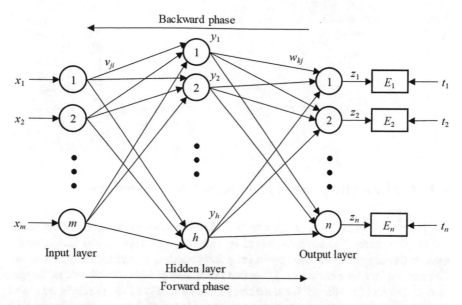

Fig. 6.1 Structure of the BP neural network with m input units, h hidden units, and n output units. v_{ji} is the weight on the connection from the input unit j to the hidden unit i, and w_{kj} is the weight on the connection from the hidden unit k to the output unit j. E_i measures the discrepancy between the expected output t_i and the actual output z_i. $E_i = (t_i - z_i)^2/2$ is generally used

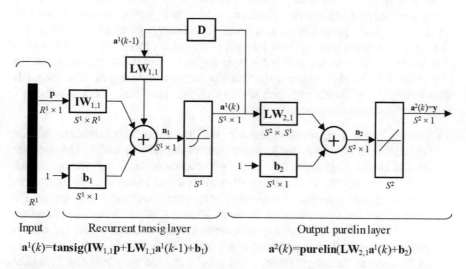

Fig. 6.2 Structure of a 2-layer Elman neural network with R^1 input units, S^1 hidden units, and S^2 output units. $\mathbf{p}$ is the input vector with R^1 elements, $\mathbf{IW}_{1,1}$ is the weight matrix from the input units to the hidden units, $\mathbf{LW}_{1,1}(S_1 \times S_1)$ is the recurrent weight matrix between the hidden units, $\mathbf{LW}_{2,1}$ is the weight matrix from hidden units to output units, $\mathbf{b}_1$ and $\mathbf{b}_2$ are bias, $\mathbf{y}$ is the expected output, **tansig** and **purelin** are activation functions for the hidden units and output units, respectively, and $\mathbf{a}^1(k)$ and $\mathbf{a}^2(k)$ are the corresponding outputs at time k

Table 6.1 Comparison of the two neural network prediction results

Methods	Test group	Average prediction accuracy
BP neural network	1	91.7%
	2	89.6%
	3	85.0%
Elman neural network	1	93.2%
	2	90.7%
	3	88.3%

6.3 CPS-Enhanced Safety Management Framework

The passenger traffic of urban rail transit is closely related to the safety of the urban rail transit system. The hazards caused by the overload of urban rail transit traffic include the triggering of large-scale crowd stampedes, terrorist attacks, epidemic diseases, and subway accidents. If we can accurately predict the subway passenger flow in different time dimensions, we can prepare a variety of safety precautions and measures in advance to avoid or mitigate these kinds of hazards. The forecasting methods of subway passenger flow lay the foundation for subway operations safety management. Furthermore, the cyber-physical system, input acquisition for forecasting models, early warning publishing and emergency control, and feedback-based continuous analysis and scheduling can be integrated into an enhanced safety management information platform, which will provide cohesive services for the emergency prevention and control of subway operations. We created a CPS-enhanced subway operations safety management system (SAFETY), whose core is the coupling of the passenger flow forecasting with the capability of the CPS (Fig. 6.3). SAFETY covers the subway safety-related topics. The model is characterized by a flexible and controllable real-time operation. The following is a detailed explanation of the proposed system.

(1) System. This is an application based on the forecasting models that are suitable for short-term (hourly to daily) predictions of the subway traffic. The accurate prediction of the passenger flow is an important requisite for subway operations safety management. Space/time series analysis and neural network models were integrated into this component. Any single method does not have a great advantage over other methods in terms of its accuracy, performance, and stability. Thus, several methods were utilized and optimized for certain prediction tasks. The application can realize the hourly to daily forecasting of the subway passenger flow; it can also query the real-time and historical passenger flow data of each station and line and thus realize the integrated management of the subway passenger flow [33].

(2) Adjust. Based on accurate prediction of the subway passenger flow, the next step is to organize, guide, and regulate the passenger flow to ensure that the passenger flow can run smoothly without large-scale centralized congestion. The subway dispatch center should follow the principle of the passenger flow

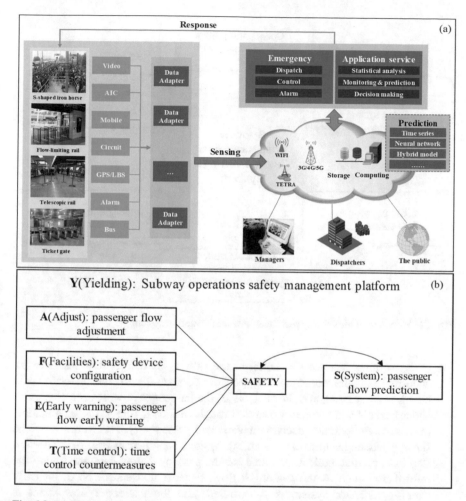

Fig. 6.3 (**a**) CPS-enhanced subway operations safety framework, (**b**) the main components of the SAFETY system

diversion priority. Regulators will formulate a large passenger flow plan at each station to ensure a smooth passenger flow based on the structure of the station and the layout of the AFC equipment, passenger flow characteristics, and equipment capabilities. For example, at present, the measures for the Beijing subway are roughly divided into three levels (limit, control, and closure) based on the passenger flow restriction guidelines.

(3) Facilities. Here, facilities refer to the related facilities and equipment that can regulate and control the subway passenger flow. The layout of these facilities will greatly affect the distribution of the inbound and outbound passenger flow, so the rational layout of facilities is crucial for subway safety guarantees [34]. The commonly used facilities for passenger flow control include S-shaped iron horses, flow-limiting rails, and telescopic rails.

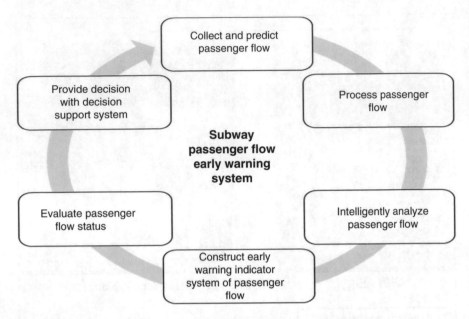

Fig. 6.4 Workflow of the subway passenger flow early warning system

(4) Early warning. The early warning of the subway passenger flow refers to the abnormal phenomenon of passenger flow compared with the historical data based on forecasting and the analysis of the passenger flow. It generates alerts about hazards and countermeasures through various media to the stakeholders (e.g., subway operation decision makers and passengers). The workflow of the subway passenger flow early warning system is shown in Fig. 6.4. The first step is to predict, collect, and analyze the passenger flow data and obtain the information such as the passenger flow, density, and speed. Next, an early warning indicator system is constructed, and the passenger flow status is evaluated based on passenger flow monitoring or prediction. The third step involves making a decision by incorporating the second step results into a decision support system, provides regulators with specific measures, and publishes safety-related information. The fourth step evaluates the implementation of the countermeasures.

(5) Time control. The volume of the subway passenger flow has a close relationship with time. The time control here mainly refers to the rational allocation of subway transport resources to guide and optimize the passenger flow. Its fundamental role is to solve the problem of the uneven distribution of the subway passenger flow in time and space, avoid the large-scale aggregation of the passenger flow, reduce passenger waiting time, and accelerate the speed of trains.

(6) Yielding. The first 5 parts comprehensively address the subway operations safety precautions based on passenger flow forecasting. However, these mea-

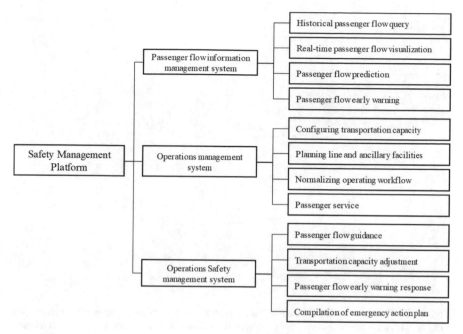

Fig. 6.5 Main components and functional structure of the subway operations safety management platform

sures are not isolated, and they need to be organically combined and applied flexibly to subway operations safety management; an integrated subway operations safety management platform is required. The platform mainly includes the passenger flow information management system, subway operations management system, and subway operations safety management system. The specific functions of each part are shown in Fig. 6.5.

The three modules of the platform complement and rely on each other to enable a rapid response to subway accidents caused by sudden changes in the passenger flow, scientifically using a variety of preventive measures to significantly improve the safety management efficiency of subway management institutions.

6.4   Application Case

6.4.1   Study Object

We used the Beijing subway system as the study object. As of December 31, 2016, Beijing had a total of 19 subway lines, with 288 operating stations, and a total mileage of 574 km, covering 11 municipal districts out of the total 18 in the city.

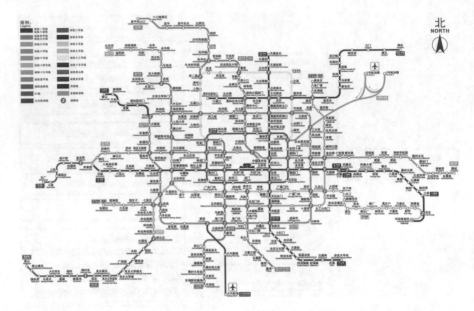

Fig. 6.6 Beijing subway map. We collected the time series data of the passenger flow from some stations. The time scales include the passenger flow data at every half an hour as well as daily information

The Beijing subway system has a 1-entrance-ticket system and a ladder fare. That is, passengers can transfer to other lines through transfer stations without exiting the subway. Passengers pay the fare according to the number of stations they travel through, from the beginning station to the terminal station. In the transfer stations, passengers can transfer to 1 or more other lines. Daily lines include both upstream and downstream directions. The Beijing subway operation map is shown in Fig. 6.6. The data used in this paper are the time-division passenger flow data, including the data taken every half an hour and daily information.

6.4.2 Time Series Prediction

In this prediction example of the time series analysis, the observations of the passenger flow every 30 min at Xizhimen station on March 4 and March 11, 2015, were used as the modeling and verification data, respectively. The observations of the passenger flow change, as shown in Fig. 6.7.

Figure 6.7 shows changes in the passenger flow at Xizhimen station on March 4th and March 11th, 2015. From the autocorrelation and partial autocorrelation plots of the first-order difference sequence, we concluded that the value of p should be 1, and the value of q should be 1 or 2, and d is the difference order 1, so

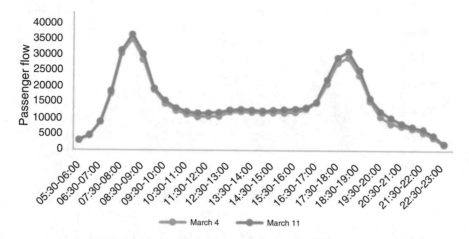

Fig. 6.7 Changes in the passenger flow at Xizhimen station on March 4th and March 11th, 2015

two possible models were obtained. The ARIMA (1,1,1) model (Eq. (6.3)) and the ARIMA (2,1,1) model (Eq. (6.4)) are shown as follows:

$$y_t = \phi_1 y_{t-1} + \varepsilon_t - \theta_1 \varepsilon_{t-1} \tag{6.3}$$

$$y_t = \varphi_1 y_{t-1} + \varphi_2 y_{t-2} + \varepsilon_t - \theta_1 \varepsilon_{t-1} \tag{6.4}$$

where all notations are as defined in Eq. (6.1).

The ARIMA modeling was performed on the above two models. The results showed that the R-squared values of the two models are greater than 0.9, and the significance levels of both are less than 0.01, which means the degree of fitting is good. The BIC value of Model 2 is smaller than that of Model 1, and the R-squared is larger than that in Model 1, so Model 2 was selected as the prediction model. The expression for Model 2 is

$$y_t = 0.98 y_{t-1} - 0.60 y_{t-2} + \varepsilon_t + 0.46 \varepsilon_{t-1} + 4379.3 \tag{6.5}$$

The fit of the model is shown in Fig. 6.8. It can be seen from the figure that the model fits very well.

With the same observations, four kinds of exponential smoothing models (the simple exponential smoothing model, Holt linear trend exponential smoothing model, Brown linear trend exponential smoothing model, and damped trend exponential smoothing model) were utilized. The obtained model parameters showed that the damped trend exponential smoothing model has the best fitting degree. Therefore, this model was used to predict and analyze the time-division passenger flow of the Xizhimen transfer station. The observations and the prediction results are shown in Fig. 6.9.

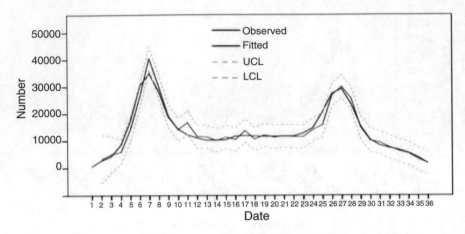

Fig. 6.8 Observations and predictions with the fitted ARIMA model at Xizhimen station

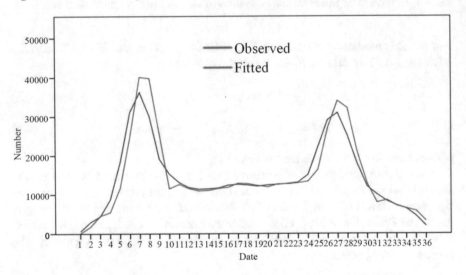

Fig. 6.9 Observations and prediction with the exponential smoothing method at Xizhimen station

6.4.3 Neural Network Prediction

The BP neural network and Elman neural network were applied to the model to predict the daily passenger flow. We examined three types of factors that should have an effect on the subway passenger flow: time, weather, and a holiday. The exploratory data analysis showed significant differences between the representative variables and passenger flow. The data for the whole year of 2015 were selected as the training groups. The input nodes are as listed in Table 6.1. The 2016 data were selected as the test groups. The data during March 7, 2016 to March 13,

2016, were selected as Test Group 1 when the weather conditions were sunny, and there were no important holidays, and the weekend factor was incorporated into the calculations. Test Group 1 focused on verifying the time impact factors. Test Group 2 used the data from July 18, 2016 to July 24, 2016, because of the continuous rainy weather in Beijing during this period, and the three days of July 19–21 experienced extreme weather conditions (from moderate to heavy rain), so these data were used to test the meteorological impact on the passenger flow. Test Group 3 included the data from February 7, 2016 to February 13, 2016, because this week had the Chinese Spring Festival Golden Week holiday, and there was no large fluctuation in the meteorological conditions, which made these times suitable for evaluating the holiday impact.

From the experiment results, we found that (1) overall, the BP neural network exhibited a high accuracy. The prediction accuracy of the three test groups exceeded 80%. (2) The prediction accuracy in special situations is lower than that in normal situations. Figure 6.10 shows the prediction accuracy of the three test groups under different training times. Test Groups 2 and 3 predicted and analyzed the daily passenger flow under extreme weather conditions and important holidays, respectively. The prediction accuracy of the two test groups is lower than that of Test Group 1. A possible cause for these results is that there were fewer training samples in the special situations, so the model could not be fully trained. Nonetheless, the accuracy was still more than 80%. If the training sample data are sufficient, the prediction accuracy will be further improved.

The same data and grouping policy were applied in the modeling of the Elman neural network. The prediction results are shown in Fig. 6.11.

The results from the Elman neural network were very similar to those exhibited by the BP neural network. From the final prediction accuracy, we also found that the Elman neural network is slightly better than BP neural network, as shown in

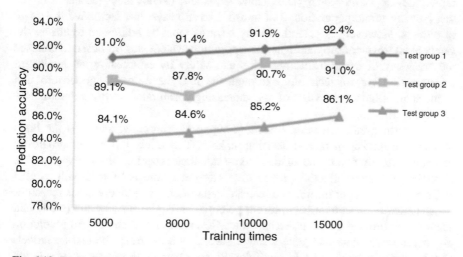

Fig. 6.10 Prediction efficiency with the BP neural network

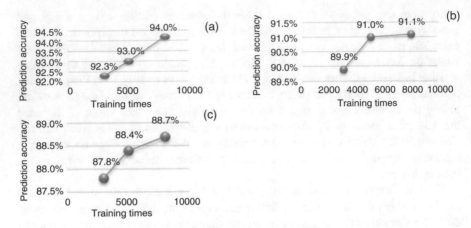

Fig. 6.11 Prediction efficiency with the Elman neural network. (**a**) Test Group 1, (**b**) Test Group 2, and (**c**) Test Group 3

Table 6.1. The training time and required training times of the Elman neural network are less than those of the BP neural network. We concluded that in predicting the subway passenger flow under the influence of multiple factors, the prediction effect of the Elman neural network is better than that of the BP neural network.

6.5 Discussion and Conclusion

The rail transit system is an important tool for travel in modern cities, and it has rapidly developed in recent years. However, it still cannot meet the needs of the fast-growing urban population. Rail transit, like subways, has facilitated the travel of citizens. However, its potential safety hazards must be addressed to ensure the safety of all travelers. An intelligent transportation system was proposed to enhance the efficiency of urban traffic operation and safety by establishing an intelligent transportation control technology through the IoT, car networking, sensors, and artificial intelligence in view of the increasingly saturated traffic infrastructure development.

Short-term prediction is an important support technology, and it is the basis of scientific management and decision-making. The safety hazards of urban rail transit are mainly due to the sudden large-scale concentration of passenger traffic. Therefore, the most effective preventative safety measure is to accurately predict and analyze passenger traffic. The relevant departments can take various measures to scientifically and effectively prevent potential safety hazards by the timely and accurate monitoring of the passenger flow. Research on the short-term prediction of the passenger flow has gone through several phases, from the classic models of time series analysis to CI-based algorithms. Overall, the newer methods are generally better than the older ones. Researchers are inclined to utilize nonlinear models and incorporate a priori information as much as possible. Because of the great success of DNN models in some fields, such as computer vision, audio, and medicine, these types of models have become popular. Among them, RNNs and their variants have been proposed to cope with sequence data, like time series and natural language; these types of models have demonstrated great potential. Some RNN models, such as LSTM, bi-directional recurrent neural network (Bi-RNN), GRU, and attention, were developed from a basic RNN (e.g., the Elman neural network) [23, 24, 35], which provided the underlying principle for them. Due to their powerful learning ability and the design of hybrid algorithms to incorporate a priori knowledge from the business domain, emerging RNN derived models should make reliable predictions of the passenger flow in urban rail transit systems.

We proposed an enhanced subway operations safety framework based on the CPS concept and prediction of the short-term passenger flow. This framework is characterized by a flexible and controllable real-time operation composed of six components: system, adjust, facilities, early warning, time control, and yielding. In this framework, the forecasting methods of the subway passenger flow are a fundamental component, and the cyber-physical system plays the roles of sensing, control, and feedback in the entire operating process; the cyber-physical system integrates with other components into an enhanced safety management information platform, which refers to the input acquisition for the forecasting models, early warning publishing and emergency control, and feedback-based re-analysis and scheduling. The implementation of the proposed framework will be able to provide integrated services for emergency prevention and the control of subway operations safety.

Acknowledgments This paper was supported in part by the National Key R&D Program of China (No 2019YFF0301300) and the National Natural Science Foundation of China (Nos. 70901047 and 7177030217). We also appreciate support for this paper from the Beijing Key Laboratory of Operation Safety of Gas, Heating, and Underground Pipelines.

References

1. X. Ma, Z. Tao, Y. Wang, H. Yu, Y. Wang, Long short-term memory neural network for traffic speed prediction using remote microwave sensor data. Transp. Res. C Emerg. Technol. **54**, 187–197 (2015)
2. A.J. Horowitz, Simplifications for single-route transit-ridership forecasting models. Transportation **12**(3), 261–275 (1984)
3. S. Zhu, X. Luo, Z. Xu, L. Ye, Seasonal streamflow forecasts using mixture-kernel GPR and advanced methods of input variable selection. Hydrol. Res. **50**(1), 200–214 (2019)
4. L.D. Galicia, R.L. Cheu, Geographic information system–system dynamics procedure for bus rapid transit ridership estimation. J. Adv. Transp. **47**(3), 266–280 (2013)
5. K.T. Azar, J. Ferreira, Integrating geographic information systems into transit ridership forecast models. J. Adv. Transp. **29**(3), 263–279 (1995)

6. J. Zhao, W. Deng, Y. Song, Y. Zhu, Analysis of Metro ridership at station level and station-to-station level in Nanjing: an approach based on direct demand models. Transportation **41**(1), 133–155 (2014)
7. J. Zhao, W. Deng, Y. Song, Y. Zhu, What influences Metro station ridership in China? Insights from Nanjing. Cities **35**, 114–124 (2013)
8. A.O. Idris, K.M. Nurul Habib, A. Shalaby, An investigation on the performances of mode shift models in transit ridership forecasting. Transp. Res. A Policy Prac. **78**, 551–565 (2015)
9. S. Chan, L. Miranda-Moreno, A station-level ridership model for the metro network in Montreal, Quebec. Can. J. Civ. Eng. **40**(3), 254–262 (2013)
10. B.D. Taylor, D. Miller, H. Iseki, C. Fink, Nature and/or nurture? Analyzing the determinants of transit ridership across US urbanized areas. Transp. Res. A Policy Prac. **43**(1), 60–77 (2009)
11. Z. Fang, X. Yang, Y. Xu, S.-L. Shaw, L. Yin, Spatiotemporal model for assessing the stability of urban human convergence and divergence patterns. Int. J. Geogr. Inf. Sci. **31**(11), 2119–2141 (2017)
12. M.G. Karlaftis, E.I. Vlahogianni, Statistical methods versus neural networks in transportation research: differences, similarities and some insights. Transp. Res. C Emerg. Technol. **19**(3), 387–399 (2011)
13. R. Xue, D.J. Sun, S. Chen, Short-term bus passenger demand prediction based on time series model and interactive multiple model approach. Discret. Dyn. Nat. Soc. **2015** (2015). https://doi.org/10.1155/2015/682390
14. X. Ma, Y.-J. Wu, Y. Wang, F. Chen, J. Liu, Mining smart card data for transit riders' travel patterns. Transp. Res. C Emerg. Technol. **36**, 1–12 (2013)
15. Z. Xiong, S. Zhong, D. Song, Z. Yu, Q. Huang, A method of fitting urban rail transit passenger flow time series. China Saf. Sci. J. **28**(11), 39–45 (2018)
16. T.-H. Tsai, C.-K. Lee, C.-H. Wei, Neural network based temporal feature models for short-term railway passenger demand forecasting. Expert Syst. Appl. **36**(2), 3728–3736 (2009)
17. Z. Xiong, J. Zheng, D. Song, S. Zhong, Q. Huang, Passenger flow prediction of urban rail transit based on deep learning methods. Smart Cities **2**(3), 371–387 (2019)
18. S. Zhu, X. Yuan, Z. Xu, X. Luo, H. Zhang, Gaussian mixture model coupled recurrent neural networks for wind speed interval forecast. Energy Convers. Manag. **198**, 111772 (2019)
19. X. Wang, K. An, L. Tang, X. Chen, Short term prediction of freeway exiting volume based on SVM and KNN. Int. J. Transp. Sci. Technol. **4**(3), 337–352 (2015)
20. Y. Sun, B. Leng, W. Guan, A novel wavelet-SVM short-time passenger flow prediction in Beijing subway system. Neurocomputing **166**, 109–121 (2015)
21. X. Jiang, L. Zhang, X.M. Chen, Short-term forecasting of high-speed rail demand: a hybrid approach combining ensemble empirical mode decomposition and gray support vector machine with real-world applications in China. Transp. Res. C Emerg. Technol. **44**, 110–127 (2014)
22. J.D. Hamilton, *Time Series Analysis*, vol. 2 (Princeton University Press, Princeton, 1994)
23. Z. Zhao, W. Chen, X. Wu, P.C. Chen, J. Liu, LSTM network: a deep learning approach for short-term traffic forecast. IET Intell. Transp. Syst. **11**(2), 68–75 (2017)
24. L. Liu, R.-C. Chen, A novel passenger flow prediction model using deep learning methods. Transp. Res. C Emerg. Technol. **84**, 74–91 (2017)
25. E.I. Vlahogianni, M.G. Karlaftis, J.C. Golias, Short-term traffic forecasting: where we are and where we're going. Transp. Res. C Emerg. Technol. **43**, 3–19 (2014)
26. G.P. Zhang, Time series forecasting using a hybrid ARIMA and neural network model. Neurocomputing **50**, 159–175 (2003)
27. E.S. Gardner Jr, Exponential smoothing: the state of the art—Part II. Int. J. Forecast. **22**(4), 637–666 (2006)
28. J. Li, J.-H. Cheng, J.-Y. Shi, F. Huang, Brief introduction of back propagation (BP) neural network algorithm and its improvement, in *Advances in Computer Science and Information Engineering* (Springer, Berlin, 2012), pp. 553–558
29. D.E. Rumelhart, G.E. Hinton, R.J. Williams, Learning representations by back-propagating errors. Nature **323**(3), 533–536 (1986)

30. K.G. Sheela, S.N. Deepa, Review on methods to fix number of hidden neurons in neural networks. Math. Probl. Eng. **2013** (2013). https://doi.org/10.1155/2013/425740
31. R.S. Toqeer, N.S. Bayindir, Speed estimation of an induction motor using Elman neural network. Neurocomputing **55**(3–4), 727–730 (2003)
32. J.L. Elman, Finding structure in time. Cogn. Sci. **14**, 179–211 (1990)
33. H.-P. Lu, Z.-Y. Sun, W.-C. Qu, Big data-driven based real-time traffic flow state identification and prediction. Discret. Dyn. Nat. Soc. **2015** (2015). https://doi.org/10.1155/2015/284906
34. X. Chen, J.W. Meaker, F.B. Zhan, Agent-based modeling and analysis of hurricane evacuation procedures for the Florida keys. Nat. Hazards **38**(3), 321 (2006)
35. F. Rui, Z. Zuo, L. Li, Using LSTM and GRU neural network methods for traffic flow prediction. Youth Academic Conference of Chinese Association of Automation, 2017

Chapter 7
A CPS-Aware Crowd Evacuation Simulation

Ze Deng

7.1 Introduction

With the continuous progress of the prosperity and development of the national economy and society, geological and man-made disasters caused very serious damages. The situation of disaster prevention and reduction is becoming more and more serious. To effectively carry out disaster prevention and mitigation, it is essential to study the special disaster risk system. The physical experiments of large scale disaster systems not only are very low feasibility but very expensive. The traditional method relies mainly on theoretical derivation and data summary, which has great limitations and blindness. With the development of computer technology, nowadays, using computer simulation experiment for disasters is an important means and method of disaster prevention and mitigation.

When unexpected events occur in large fields with large-scale crowd, the crowd evacuation is particularly important. Research on safety evacuation of large-scale crowd has been lasted for a long time. Some typical simulation models have been proposed (e.g., cellular automata model, multi-agent model, hydrodynamic model [6], and social force model [5]). However, these simulation models focus on a closed simulation environment. That means the simulations cannot receive information and data from surroundings. Nevertheless, in real life, with the development of IoT and CPS, the crowd simulations need to dynamically adjust and correct their simulation steps, according to inputting data from sensors or analyzed information. Thus, there is a pressing need for a new approach for crowd simulation in an open environment

Z. Deng (✉)
School of Computer Science, China University of Geosciences, Wuhan, Hubei, P.R. China

Hubei Key Laboratory of Intelligent Geo-Information Processing, China University of Geosciences (Wuhan), Wuhan, P.R. China

© Springer Nature Switzerland AG 2020
S. Hu, B. Yu (eds.), *Big Data Analytics for Cyber-Physical Systems*,
https://doi.org/10.1007/978-3-030-43494-6_7

171

so that data or information can input the simulation systems to change the behavior of simulation systems for better simulation results.

"Cyber-physical system" (CPS) has been introduced by Helen Gill at the National Science Foundation in the USA in 2006 [9]. Based on the definition in [9], "cyber" refers to virtual computational and analytical process that has been well developed to control and manipulate the physical entity to perform predefined tasks. CPS is a typical feedback system according to input information. Therefore, the idea behind CPS can well deal with the issue of current crowd simulation model. Based on the inspiration of CPS, in this paper, I propose a CPS-aware crowd simulation framework. The framework can help a simulation system to conduct and correct "better" results. The framework mainly consists of two components: (1) the simulation mode and (2) the feedback mode. The simulation mode executes data-driven micro modeling and simulation. Then, the feedback mode receives and clusters past trajectories from crowd moving objects during the simulation and guides the simulation mode based on the clustering results.

The remainder of this paper is organized as follows: Section 7.2 discusses work relating to crowd simulation. Section 7.3 introduces the CPS-aware simulation framework. Section 7.4 presents the case study of the proposed approaches. Section 7.5 concludes with a summary.

7.2 Related Work

This section describes the most salient works along crowd simulation. The models about crowd simulation systems can be roughly classified into two categories, i.e., macro method and micro method. The macro modeling and simulation method is a method of modeling groups at the individual level. The basic idea is to set a certain number of groups described as individuals [1]. In the micro modeling and simulation methods, the interaction between individual and individual and the interaction between individuals and environment are directly reflected by the simulation process.

7.2.1 Micro Modeling Methods

According to the difference between individual complexity and autonomy, I introduce two typical micro models: the particle system [1] and multi-agent system [14]. Particle system is the simplest method of micro modeling. In a particle system, each individual of a group is regarded as a physical particle. Each particle has the properties of mass, initial velocity, initial velocity direction, and so on. The motion of a particle follows the classical laws of motion such as Newton's law of motion and the law of conservation of momentum. In the n-dimensional Euclidean space where the particles are located, the motion of a particle follows the formula (7.1):

$$\begin{cases} \frac{d}{dt}x=v_x \\ \frac{d}{dt}(m \times v_x)=F_x \end{cases} \tag{7.1}$$

where m represents the quality of a particle, t is time, and x indicates the distance that particles move in the current dimension. Meanwhile, v_x reflects the instantaneous rate of particles on the current dimension, F_x is the component of the force acting on the particle in the current dimension. Based on the formula (7.1), the interaction between particles and the interaction between particles and environment are mainly achieved by collision and field force.

The multi-agent system gives the individual simple attributes and decision logic. The crowd modeling and simulation method based on multi-agent system [14] describes the individual of a group as an agent and constructs a set of attributes and decision logic to realize the representation of individual behavior. Through the decision-making and execution process of agents, the process of crowd development and evolution is simulated.

7.2.2   Macro Modeling Methods

The macro modeling and simulation method is a method of modeling groups at the level of groups. The basic idea is to describe the group itself or its movement variation as a physically continuous and variable whole [8]. The representative methods are fluid mechanics model [11] and macroscopic field model [13].

The fluid mechanics model describes the flow of crowd as gas or fluid to apply the model of gas dynamics and fluid mechanics to pedestrians. The description of pedestrian behavior using a gas dynamic model was first proposed by Henderson in 1971 [7]. Henderson gives the probability distribution formula of pedestrian forward velocity by using the Maxwell–Boltzmann distribution in thermodynamics. Based on the Boltzmann equation in 1998, Helbing proposed a more realistic model of pedestrian behavior of fluid [4]. The complexity and computational complexity of this kind of simulation model are relatively low, and it is suitable for the study of traffic state of road network. Another typical macro model is macroscopic field model [13]. Different from fluid model, the modeling object of the macroscopic field model is not the crowd itself, but the movement trend of the crowd. The motion trend of the population in the macro is described as a vector field in the whole environment, and the evolution of the population is described by the distribution, direction, and intensity of the vector field. In the macroscopic field modeling and simulation, the artificial potential field [10] is more representative. In this method, the whole environment of population distribution is described as a surface with height difference. In the course of the crowd movement, the height of the area tending to be concentrated is relatively low in the corresponding position of the surface so that the potential energy is relatively small. On the other hand, people tend to stay away from the area, and the height of the corresponding position in the surface is relatively high. Correspondingly, potential energy is relatively large.

However, these mentioned simulation models focus on a closed simulation environment. That means the simulations cannot receive information and data from surroundings. In this paper, I propose a CPS-aware crowd simulation framework. The framework can help simulation system to conduct and correct "better" results.

7.3   The CPS-Aware Simulation Framework

The proposed framework mainly consists of two components: a simulation mode and a feedback mode. Figure 7.1 illustrates the framework. As we can see, the simulation mode executes crowd simulation. Then, the feedback mode receives past trajectories from crowd moving objects during the simulation and gains trajectories clustering results to guide the following simulation rounds.

7.3.1   Crowd Simulation

In this section, I choose a multi-agent system to simulate a large-scale crowd. The reason for choosing a micro model to simulate crowd is that I want to collect trajectories for all moving objects. In this model, the simulation procedure includes the following steps:

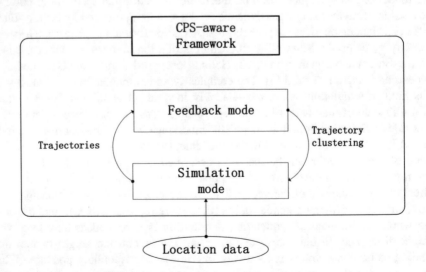

Fig. 7.1 The CPS-aware simulation framework

1. Initialization: initialize the simulation based on inputting location data.
2. Mesh generation: the plane of the simulation map is divided into uniform grid information, and the single grid corresponds to the single cell. According to China national standard GB-10000-88, I calculate personal projection area. A single cell roughly corresponds to the range of 0.4 m × 0.4 m
3. Making grid attributes: Each grid attribute is different. Some grid attributes are obstacles, some grid attributes are agent, there are guidance signs and exits, and so on. Grid diagrams are automatic generation through programs. Agent can be added either by random initialization or by changing a single grid attribute. Other attributes are determined by map manipulation during map design. Meanwhile, I can store the designed map scenes for reuse.
4. Definition of personal characteristics. In our setting, I only consider the degree of physical strength of evacuees. It can be divided into four types: youth, old age, disability, and children.
5. Possible direction of movement. The model uses the Moore type neighborhood of the CA model, and the evacuation personnel have 8 possible moving directions.
6. Agent mobile decision. Each agent mobile decision combines with the surrounding environment and itself (see formula (7.2)).

$$Des\_Agent = < Mac_t, Loc_t, Sel_t >$$ (7.2)

where Mac_t is macro scene information at t time, Loc_t is agent motion at t time, and Sel_t presents self-characteristics.

7. Attribute definition for each agent. The definition sees formula

$$Agent_i = \begin{bmatrix} x \\ y \\ H \\ Dir \\ V \\ mV \\ D \\ Time \\ Sight \\ W_{dir} \\ W_{pos} \\ W_{den} \end{bmatrix}$$ (7.3)

where x and y are row and column coordinates in grid. H, Dir, and V represent current health degree, direction, and speed, respectively. mV is the upper bound of speed. D represents the crowd density at current location. Time is the total time to escape. W_{dir}, W_{pos}, and W_{den} reflect the attraction weights of direction, position, and density.

7.3.2 The Feedback Analysis

The basic definition of trajectory is composed of timestamp and the location of the corresponding timestamp in the two dimensional space. Comparing the agent movement of our simulation system, $agent_i$ in the time slice t must exist at a point in the simulation map (x, y). Therefore, the movement of $agent_i$ from the start position to the exit position can be defined as $T_i = \{(z_1, t_1), (z_2, t_2), \ldots, (z_j, t_j)\}$. The movement track of each agent is recorded during the simulation. A set of trajectory data sets D can be obtained at the end of the simulation. Thus, the trajectory clustering analysis based on density clustering method can be carried out on trajectory data set D. Based on the clustering results, the feedback mode can classify the trajectory data set D according to the clustering ID of each trajectory to guide the following simulations. As we can see, the important part in the feedback mode is to clustering the trajectories from past simulations. In the following sections, I begin to introduce how to effectively cluster trajectories. I used Tra-POPTICS that is a scalable trajectory clustering algorithm in [2]. I first introduce the POPTICS algorithm. Then, the details of the proposed Tra-POPTICS are described including the distance function, the description of the Tra-POPTICS algorithm, and the optimization of Tra-POPTICS using a trajectory indexing technique.

7.3.2.1 POPTICS

OPTICS is a density-based clustering algorithm and can detect meaningful clusters in data of varying density by producing a linear order of points such that points which are spatially closest become neighbors in order [12]. OPTICS starts with adding an arbitrary point of a cluster to the order list and then iteratively expands the cluster by adding a point within ε-neighborhood of a point in the cluster which is also closest to any of the already selected points. The process repeats for remaining clusters. Meanwhile, OPTICS also computes the reachability distance for each point. The clusters for any clustering distance ε' $(\varepsilon' < \varepsilon)$ can be extracted based on the computed order and reachability distances.

POPTICS [12] is a scalable OPTICS algorithm using graph algorithms concepts. POPTICS exploits the similarities between OPTICS and MST algorithm to break the sequential access of data points in the classical OPTICS algorithm. The POPTICS algorithm is designed based on the observation that Prim's approach to continuously increase one edge to the MST at a time is very similar to the OPTICS algorithm. The vertices of MST are analogous to the points of OPTICS and the edge weights of MST are analogous to the reachability distances of OPTICS. Therefore, POPTICS expended MST to support OPTICS. As a result, POPTICS can divide the whole data set into local data subsets and let each process core or machine compute local MSTs through running the OPTICS based on its local data points in a distributed way. Then, a parallel merge of the local MSTs is performed to obtain the global MST. The clusters for any clustering distance ε' can be extracted from the global MST. As such, POPTICS can obtain a great scalability through avoiding to sequentially compute the reachability distances and order for all points compared to classical OPTICS. However, POPTICS aims for point data. Therefore, in the following subsection, I will introduce how to drive POPTICS to support trajectory data.

7.3.2.2 Tra-POPTICS: POPTICS for Clustering Trajectory Data

To successfully support the trajectory clustering using POPTICS, firstly a distance function for trajectory data [3] is used for measuring the dissimilarity between two trajectories. Then, I introduce the POPTICS algorithm for trajectory data called Tra-POPTICS.

In POPTICS, the distance function measures the Euclidean distance between two points. In our setting, the distance function needs to measure the spatiotemporal distance between two trajectories. Therefore, I apply a distance metric of spatiotemporal dissimilarity between trajectories [3] in our scenario. For simplification, I assume that the moving objects move linearly with time in a 2D plane in this paper.

Concretely speaking, given two trajectories T_1 and T_2, the spatiotemporal distance between T_1 and T_2 Dist(T_1, T_2) is approximately computed as: Dist(T_1, T_2)

$$\approx \sum_{i=1}^{n-1} \left((D_{T_1, T_2}(t_k) + D_{T_1, T_2}(t_{k+1})) \times (t_{k+1} - t_k) \right) \tag{7.4}$$

where $[t_1, t_n]$ is the common time period between T_1 and T_2 and t_i is a timestamp for computing distance. $D_{T_1, T_2}(t)$ is the Euclidean distance with time. Noted that linear interpolation is applied for formula (7.1) because different trajectory is represented by a collection of discrete points with various sampling rates. $D_{T_1, T_2}(t)$ is computed based on the definition in [3]:

$$D_{T_1, T_2}(t) = \sqrt{a \times t^2 + b \times t + c} \tag{7.5}$$

where let the line segments for T_1 and T_2 be $(p_{t_k}, p_{t_{k+1}})$ and $(q_{t_k}, q_{t_{k+1}})$ at timestamp t_k and t_{k+1} with linear interpolation, the values of a, b, and c in formula (7.5) are computed as follows:

$$a = \frac{A}{(t_{k+1} - t_k)^2}$$

$$b = \left(\frac{B}{t_{k+1} - t_k} - \frac{2 \times A \times t_{k+1}}{(t_{k+1} - t_k)^2} \right)$$

$$c = \frac{A \times t_k^2}{(t_{k+1} - t_k)^2} - \frac{B \times t_k}{t_{k+1} - t_k} + C$$

With

$$A = (q_{t_{k+1}}.x - q_{t_k}.x - p_{t_{k+1}}.x + p_{t_k}.x)^2$$
$$+ (q_{t_{k+1}}.y - q_{t_k}.y - p_{t_{k+1}}.y + p_{t_k}.y)^2$$
$$B = 2((q_{t_{k+1}}.x - q_{t_k}.x - p_{t_{k+1}}.x + p_{t_k}.x)(q_{t_k}.x - p_{t_k}.x)$$
$$+ (q_{t_{k+1}}.y - q_{t_k}.y - p_{t_{k+1}}.y + p_{t_k}.y)(q_{t_k}.y - p_{t_k}.y))$$
$$C = (q_{t_k}.x - p_{t_k}.x)^2 + (q_{t_k}.y - p_{t_k}.y)^2$$

In the formulas for computing the values of A, B, and C, $p.x$, $p.y$ and $q.x$ and $q.y$ are the coordinates of points p and q in the x and y axis.

Using the above-mentioned distance function $Dist(T_1, T_2)$ to compute the spatiotemporal distance of any pair of trajectories, I can design the algorithm of POPTICS for clustering trajectory data (Tra-POPTICS). Like POPTICS, Tra-POPTICS can be implemented based on shared memory and distributed memory. In this paper, I focus on Tra-POPTICS based on shared memory. Such that, the Tra-POPTICS algorithm is presented in Algorithm 1. As we can see, the Tra-POPTICS algorithm consists of three steps:

Step1: Location MST computing (Line 2–23): each CPU thread processes a local disjointed subset of trajectory data. For each CPU thread, it first finds out ε-neighbors of each trajectory with *FindNeighbor* function. The *FindNeighbor* function is presented in Algorithm 2. In Algorithm 2, I simply linearly scan each item in the subset with the two steps: interpolating two trajectories within the common time period (Line 4 and 5 in Algorithm 2) and measuring the distance between two trajectories using the distance function presented in formula (7.1) (Line 6 in Algorithm 2). After getting neighbors, each CPU thread computes the core distance of each trajectory (Line 8) and then local MST (Line 9–23). In this step, since p threads process the local data in parallel and p is the maximum active threads provided with CPU, the computational cost is $O(n' \times$ runtime of an ε-neighborhood query), where n' is the number of trajectories in the local subset. According to Algorithm 2, the query processing requires a linear scan of the entire subset. Consequently, the time complexity of the step 1 is $O(n'^2)$. Let n be the number of all trajectories, I can get $n' = n/p$ because the entire data set is equally assigned p threads. Such that, the complexity is $O((n/p)^2)$.

Step 2: Generating the global MST (Line 24–32) and step 3: extracting clusters from the global MST (Line 33–39) is similar to the one in POPTICS. The time complexity in both steps is $O(n)$. Therefore, the computing complexity of Tra-POPTICS is $O((n/p)^2 + 2n)$.

Algorithm 1: Tra-POPTICS based on shared memory

---

1 **Clustering_Procedure(D, p, ε, ε', minNumofTrs, CID)**/* Input : D is a set
     of trajectory data, p is the number of CPU threads, ε
     means the spatiotemporal distance bound for clustering,ε'
     is the clustering distance, minNumofTrs is the minimum
     number of trajectories in one cluster   */
     /* Output: CID clusters in CID   */
2 Divide D into p equal-sized subsets $D_1,D_2,\ldots,D_p$. Each subset is assigned to one CPU
     thread.
3 set a queue Q shared among all CPU treads for generating the global MST
4 **for** *each CPU thread t_i ($1\leq i\leq p$)* **in parallel do**
5   **for** *each trajectory element $tr \in D_i$ ($1\leq i\leq p$)* **do**
6   mark tr as processed
7   Ns $\leftarrow$ FindNeighbors(tr, D_i, ε)
8   GenCoreDistance(tr, Ns, ε, minNumofTrs)
9   **if** *tr.coreDistance $\neq$ NULL* **then**
10   Update(tr, Ns, P_i)
11   **while** *$P_i \neq$ empty* **do**
12   (tr1, tr, w) $\leftarrow$ findMin(P_i)
13   insert (tr1, tr, w) into Q
14   **if** *tr1 $\in D_i$* **then**
15   mark tr1 as processed
16   Ns' $\leftarrow$ FindNeighbors(tr1, D_i, ε)
17   GenCoreDistance(tr1, Ns', ε, minNumofTrs)
18   **if** *tr1.coreDistance $\neq$ NULL* **then** Update(tr1, Ns', P_i)
19   ;
20   **end**
21   **end**
22   **end**
23 **end**
24 **end**
     /* Merge all local results to get the global MST   */
25 Set a empty MST T
26 **for** *each trajectory $tr \in D$* **in parallel do**
27   tr.parent $\leftarrow$ tr
28 **end**
29 **while** *Q $\neq$ empty* **do**
30   (t1, t2, w) $\leftarrow$ findMin(Q)
31   **if** *t1.parent $\neq$ t2.parent* **then** Union(t1, t2)
32   T $\leftarrow$ T $\cup$ (t1, t2, w) ;
33 **end**
     /* clustering from MST with any ε' ($< \varepsilon$)   */
34 **for** *each trajectory $tr \in D$* **do**
35   CID[tr.ID] $\leftarrow$ tr.ID
36 **end**
37 **for** *each edge (tr1, tr2, w) $\in T$* **do**
38   **if** *$w \leq \varepsilon'$* **then** CID[tr1.ID]$\leftarrow$ tr2.ID or CID[tr2.ID]$\leftarrow$ tr1.ID ;
39 **end**
40 return CID

Algorithm 2: Find trajectory Neighbors

---

1 **FindNeighbors(tr, D, ε)**/$*$ Input : tr is a trajectory, D is a set of
       trajectory data for neighborhood candidates, ε means the
       spatiotemporal distance bound for clustering, $*$/
   /$*$ Output: RS is a set of ε-neighbors $*$/
2 RS $\leftarrow \emptyset$
3 **for** *each trajectory* $d_i \in D$ *($1 \leq i \leq |D|$)* **do**
4 | d_i' $\leftarrow$ Interpolate d_i with time period [max(d_i.startTS, tr.startTS), min(d_i.endTS,
       tr.endTS)]
5 | tr' $\leftarrow$ Interpolate tr with time period [max(d_i.startTS, tr.startTS), min(d_i.endTS,
       tr.endTS)]
6 | ds $\leftarrow$ Dist(d_i', tr') // Dist function see formula (7.1)
7 | **if** $ds \leq \varepsilon$ **then** RS $\leftarrow$ RS $\cup$ {d_i }
8 | ;
9 **end**
10 return RS

7.4 Case Study

This section presents the result and analysis of related case study of simulation model. First, I evaluate the correctness of the simulation system by using a crowd evacuation case. Then, the simulation trajectory data are clustered and the simulation path optimization is tested based on clustering results. All experiments were executed on one computer and the configurations are presented in Table 7.1.

7.4.1 Simulation for Crowd Evacuation

The experimental case procedure is as follows:

- Step 1: Initialization map information. Table 7.2 shows that parameters initialize a 100×100 map, and randomly initialize 4 exports, the map information into the file.
- Step 2: According to the simulation method in Sect. 7.3.1, the results are calculated and output.
- Step 3. The simulation results are illustrated in Fig. 7.2.

Table 7.1 Configurations of the computer

Specifications of CPU platforms	Computer
OS	Windows 7 64
CPU	i7-4790 (3.60GHz, 4 cores)
Memory	32GB DDR3

Table 7.2 Initializing map information

Map width	100
Map length	100
Number of exports	4

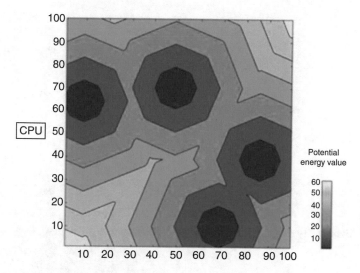

Fig. 7.2 Simulation result

In Fig. 7.2, the contour of the image is drawn by contour lines. The color is yellow, which means that the potential energy value is higher, the color is darker, and the potential energy value of blue is smaller. The four deep blue sections are four exits, and the lower the potential energy is near the exit. The simulation result conforms to the actual crowd evacuation.

7.4.2 Crowd Evacuation Trajectory Clustering for Simulation

I first collect crowd trajectories in simulation system based on the simulation case in Sect. 7.4.1. The output of each time slice agent is recorded after moving. There are 542 agent in the simulation scene, so there are 542 track data, and Fig. 7.3 selects the trajectory data that the agent sequence number can be divided by 30. Meanwhile, all trajectories are shown in Fig. 7.4.

Then, the path choice for the next simulation round is optimized through clustering analysis of the 542 crowd trajectories. The parameters of Tra-POPTICS algorithm are shown in Table 7.3. Output is trajectory data sets with different clustering numbers. After analyzing the trajectory data set of clustering results, the simulation tracks of different categories are found, and the simulation scene is analyzed concretely.

Fig. 7.3 Partial trajectory
data

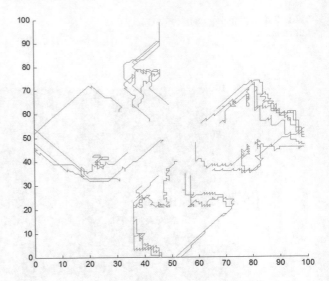

Fig. 7.4 All trajectory data

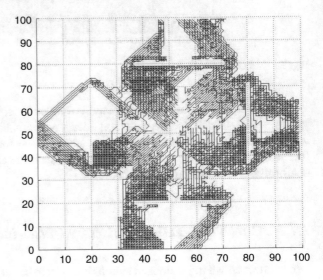

Table 7.3 Parameters

ε	**MinTra**
350	3

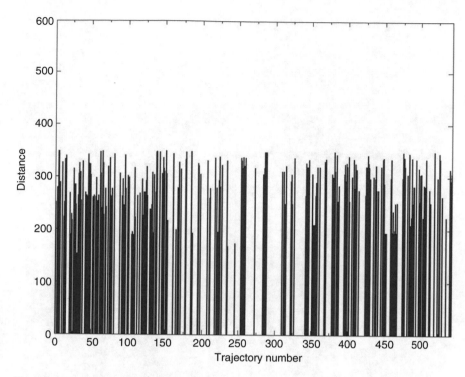

Fig. 7.5 Trajectory clustering results

Experimental results in Fig. 7.5 show that the abscissa is the trajectory number, and the ordinate is the reachable distance value. The histogram of the graph well shows the reachable distance of different trajectories.

Figure 7.6a–d reflects the trajectory clustering results with $\epsilon = 260$, $\epsilon = 280$, $\epsilon = 300$, and $\epsilon = 320$, respectively. When the extraction factor ϵ is small, the number of trajectories in clusters is less while When the extraction factor ϵ is large, there are more trajectories in the class cluster. However, four different extraction factors are basically consistent with the results of trajectory clustering. There are eight main routes of distribution. Comparing the original simulation trajectory of figure, I can get a better path in the simulation map. The path distribution is basically extracted with clustering. Through the clustering analysis of the simulation crowd trajectory, I can extract the key trajectory data from the complex trajectory data set, and these key trajectory data can well guide the simulation path planning.

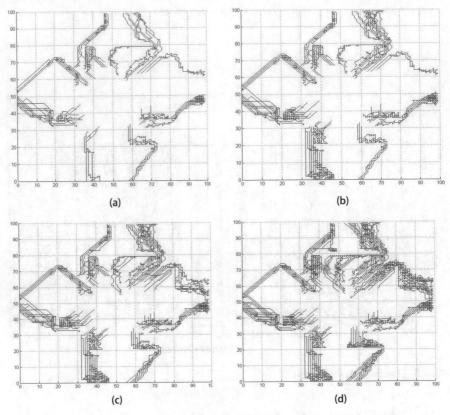

Fig. 7.6 Trajectory extraction. (**a**) $\epsilon = 260$, (**b**) $\epsilon = 280$, (**c**) $\epsilon = 300$, (**d**) and $\epsilon = 320$

7.5 Conclusion

For dealing with unexpected events in disaster cases with large-scale crowd, crowd evacuation is particularly important. Research on safety evacuation of large-scale crowd has been lasted for a long time. However, these simulation models focus on a closed simulation environment. That means the simulations cannot receive information and data from surroundings. Nevertheless, in real life, with the development of IoT and CPS, the crowd simulations need to dynamically adjust and correct their simulation steps, according to inputting data from sensors or analyzed information. Therefore, in this study, I propose a CPS-aware crowd simulation framework. The framework can help simulation system to conduct and correct "better" results. The framework mainly consists of two components: (1) the simulation mode and (2) the feedback mode. The experimental results show that I can get better evacuation paths with our crowd simulation framework.

Acknowledgments This paper was supported in part by the National Natural Science Foundation of China (No. U1711266) and the China Postdoctoral Science Foundation (2014M552112).

References

1. E. Bouvier, E. Cohen, L. Najman, From crowd simulation to airbag deployment: particle systems, a new paradigm of simulation. J. Electron. Imaging **6**, 94–107 (1997)
2. Z. Deng, Y. Hu, M. Zhu, X. Huang, B. Du, A scalable and fast OPTICS for clustering trajectory big data. Clust. Comput. **18**, 549–562 (2015)
3. E. Frentzos, K. Gratsias, Y. Theodoridis, Index-based most similar trajectory search. In: IEEE International Conference on Data Engineering, 2007, pp. 816–825
4. D. Helbing, A fluid dynamic model for the movement of pedestrians. Complex Syst. **6**, 391–415 (1992)
5. D. Helbing, I. Farkas, T. Vicsek, Simulating dynamical features of escape panic. Nature **407**, 487–490 (2000)
6. D. Helbing, I.J. Farkas, P. Molnar et al., Simulation of pedestrian crowds in normal and evacuation situations, in *Pedestrian and Evacuation Dynamics*, ed. by M. Schreckenberg, S.D. Sharma (Springer, Berlin, 2002), pp. 21–58
7. L.F. Henderson, The statistics of crowd fluids. Nature **229**, 381–383 (1998)
8. R.L. Hughes, The flow of human crowds. Annu. Rev. Fluid Mech. **35**, 169–182 (2003)
9. J. Lee, C. Jin, B. Bagheri, Cyber physical systems for predictive production systems. Prod. Manag. **11**, 155–165 (2017)
10. I. Nagy, W.K. Fung, P. Baranyi, Neuro-fuzzy based vector field model: an unified representation for mobile robot guiding styles. In: IEEE International Conference on Systems, 2000, pp. 3538–3543
11. S. Patil, J. van den Berg, S. Curtis et al., Directing crowd simulations using navigation fields. IEEE Trans. Vis. Comput. Graph. **17**, 244–254 (2011)
12. M.M.A. Patwary, D. Palsetia, A. Agrawal, W. Keng Liao, F. Manne, A. Choudhary, Scalable parallel optics data clustering using graph algorithmic techniques. In: The International Conference for High Performance Computing, Networking, Storage and Analysis, 2013, pp. 49:1–49:12
13. A. Treuille, S. Cooper, Z. Popović, Continuum crowds. In: United States: Association for Computing Machinery, 2006, pp. 1160–1168
14. N. Zarboutis, N. Marmaras, Design of formative evacuation plans using agent-based simulation. Saf. Sci. **45**, 920–940 (2007)

Chapter 8
Smart Building Sensor Drift Calibration

Tinghuan Chen, Bingqing Lin, Hao Geng, and Bei Yu

8.1 Introduction

In modern smart building, the temperature measurement is a key step for smart temperature management implemented by a cyber-physical system (CPS) [1, 2]. CPS is a complex, heterogeneous distributed system with seamlessly integrated and closely interacted cyber components (e.g., sensors, sink nodes, control centers, and actuators) and physical processes (e.g., temperature) [3]. As shown in Fig. 8.1, the physical world is sensed by corresponding sensors and the acquired data is sent to a sink node or control center. Then the sink node or control center will send an instruction to actuators to control the physical world after the data is analyzed. In smart building, the in-building temperatures are monitored by several spatially distributed and immovable temperature sensors.

T. Chen · H. Geng · B. Yu (✉)
CSE Department, Chinese University of Hong Kong, Hong Kong, China
e-mail: thchen@cse.cuhk.edu.hk; hgeng@cse.cuhk.edu.hk; byu@cse.cuhk.edu.hk

B. Lin
Shenzhen University, Shenzhen, China
e-mail: bqlin@szu.edu.cn

© Springer Nature Switzerland AG 2020
S. Hu, B. Yu (eds.), *Big Data Analytics for Cyber-Physical Systems*,
https://doi.org/10.1007/978-3-030-43494-6_8

187

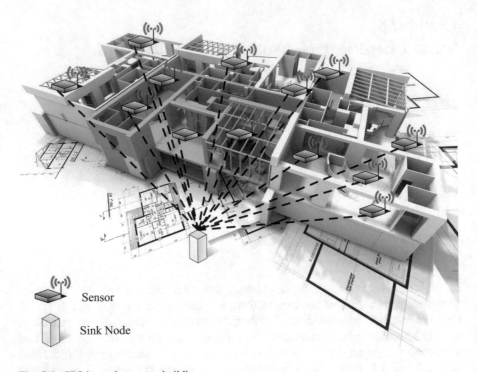

Sensor

Sink Node

Fig. 8.1 CPS in modern smart building

Although advanced technologies in the semiconductor industry and micro-electromechanical systems are developed in recent years, in practice, sensors outputs exist errors, which are one of the major barriers to the use of sensor networks. There are three main types of errors: gain, drift, and noise [4]. Compared with gain and noise, the sensor drift is considered with vital importance since it has significantly negative effect on measurement accuracy [5]. Although sensors with high accuracy can be deployed, these sensors always have expensive price. As shown in Fig. 8.2a, the temperature sensor AD590JH with ±0.5 °C accuracy is sold at more than tenfold price of TMP100 with ±2 °C accuracy.

Sensor drift calibration has been studied in many literatures. Without further assumption, calibration cannot be performed. In [7–9], at most one sensor is assumed to have an unknown drift, which is estimated by Kalman filter. In practice, this assumption is hard to satisfy. Therefore, the calibration problem is naturally studied extensively to be a sparse reconstruction problem, where a sparse set of sensors are assumed to have significant drifts. These drift calibration works mainly depend on subspace prior, which is first proposed by Balzano and Nowak to perform calibration when variational sources are over-sampled by sensors [10]. The projection matrix is obtained by singular value decomposition (SVD) [10, 11]. In [11], Wang et.al. adopt temporal sparse Bayesian learning (TSBL) [12] to calibrate time-variant and incremental drifts for the sparse set of sensors. However, due to the

Part Number	Temp. Range	Accuracy	Price [12]
SMT172	−45 ∼ 130 °C	±0.25 °C	$ 35.13
AD590JH	−50 ∼ 150 °C	±0.5 °C	$ 17.91
TMP100	−55 ∼ 125 °C	±2.0 °C	$ 1.79
MCP9509	−40 ∼ 125 °C	±4.5 °C	$ 0.88
LM335A	−40 ∼ 100 °C	±5.0 °C	$ 0.75

(a)

(b)

(c)

Fig. 8.2 (a) Comparison of different temperature sensors; (b) Sensor MCP9509; (c) Sensor LM335A

sparsity assumption, not all sensors can be calibrated. In addition, since observation matrix is directly determined by drift-free measurement, the method cannot calibrate drifts if signals lie in time-variant subspace.

Very recently, in order to calibrate all sensors, Ling and Strohmer presented three models, which are formulated as bilinear inverse problems [13]. However, these models heavily rely on the partial information about the sensing matrix. For the temperature sensor calibration in a smart building, the sensing matrix depends on weather, the position of sensors, and parameters of the building, e.g., material characteristics, geometry, and equipment power per area [1, 2, 14]. In practice, it is hard to obtain these complex and tedious information. As a result, these models cannot be directly used to calibrate temperature sensors in a smart building.

In this paper, we focus on the temperature sensor drift calibration. Several low-cost sensors with low accuracy are deployed to sense in-building temperatures (see Fig. 8.2b,c). Unlike traditional arts, we build a sensor spatial correlation model whose coefficients only depend on measurements, and we assume that all sensors have drifts. Our model coefficients are optimally determined by statistically extracting prior information from drift-free measurement model coefficients and *maximum-a-posteriori* (MAP) estimation. As a result, our proposed sensor drift calibration framework allows that the signals lie in time-variant subspace. MAP estimation is formulated as a non-convex problem with three hyper-parameters. We propose an alternating-based optimization algorithm to handle the non-convex formulation. Cross-validation and *expectation-maximization* (EM) with Gibbs sampling are adopted to determine hyper-parameters, respectively.

Experimental results show that on benchmarks simulated from `EnergyPlus`, compared with state-of-the-art method, the proposed framework with EM can achieve a better trade-off between accuracy and runtime.

The rest of this paper is organized as follows. In Sect. 8.2, we provide a problem formulation about sensor drift calibration and broadly introduce our proposed whole flow. In Sect. 8.3, we build a drift calibration model based on sensor spatial correlation and deliver mathematical formulation with three hyper-parameters.

In Sect. 8.4, we propose a more efficient method to handle the mathematical formulation. In Sect. 8.5, three hyper-parameters are determined by cross-validation and EM with Gibbs sampling, respectively. Section 8.6 presents experimental results with comparison and discussion, followed by conclusion in Sect. 8.7.

8.2 Preliminary

8.2.1 Problem Formulation

Several low-cost sensors are deployed to sense in-building temperatures. Due to a slow-aging effect, all sensors have unknown time-invariant drifts. As shown in Fig. 8.3, unlike communication channels [12], for a sensor signal to be output, e.g., current, it is contaminated by a time-invariant drift. In order to achieve high-accurate measurements, drifts need to be estimated and calibrated. Specifically, the mean absolute percent error (MAPE) is used to evaluate drift calibration accuracy.

Based on the above description, we define the sensor drift calibration problem as follows.

Problem 1 (Sensor Drift Calibration) Given the measurement values sensed by all sensors during several time-instants, drifts will be accurately estimated and calibrated.

8.2.2 Overall Flow

The overall flow of our proposed sensor drift calibration is shown in Fig. 8.4, which consists of three parts: model optimization, cross-validation, and EM with Gibbs sampling.

After drift-free measurements model coefficients and several temperature measurements with drifts are input, an alternating-based optimization algorithm is proposed to handle sensor drift calibration formulation in model optimization. In addition, cross-validation and EM with Gibbs sampling are adopted to induce

Fig. 8.3 Drift
vs. temperature [15]

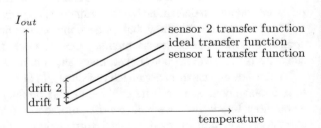

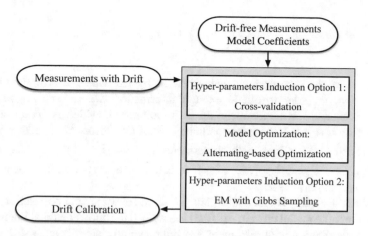

Fig. 8.4 The proposed sensor drift calibration flow

hyper-parameters, respectively. By using the proposed sensor drift calibration, it is expected to accurately calibrate sensor drifts.

8.3 Mathematical Formulation

We assume that n sensors are deployed to sense in-building temperatures. During a short time after new sensors are deployed, the drift is assumed to be insignificant. Furthermore, like [11], we assume all sensors are drift-free during m_0 initial time-instants. Due to over-sampling, as illustrated in [10, 11], signals measured by sensors lie in a low dimensional subspace. Furthermore, in a smart building, all actual temperatures measured by sensors have a high correlation, for example, the dense deployment of sensors. Therefore, we build a linear model among all actual temperatures as follows:

$$x_i^{(k)} \approx \sum_{j=1, j \neq i}^{n} a_{i,j} x_j^{(k)} + a_{i,0}, \qquad k = 1, 2, \ldots, m_0, \qquad (8.1)$$

where $x_i^{(k)}$ is the ground-truth temperature sensed by ith sensor at kth time-instant. $a_{i,j}$ is the drift-free model coefficient. We define $\mathbf{x} = [x_1^{(1)}, x_2^{(1)}, \ldots, x_n^{(1)}, \ldots, x_n^{(m_0)}]^{\top}$, $\mathbf{a}_i = [a_{i,0}, \ldots, a_{i,i-1}, a_{i,i+1}, \ldots, a_{i,n}]^{\top} \in \mathbb{R}^n$, $\mathbf{a} = [\mathbf{a}_1^{\top}, \mathbf{a}_2^{\top}, \ldots, \mathbf{a}_n^{\top}]^{\top} \in \mathbb{R}^{n^2}$.

Due to a slow-aging effect, all sensors have unknown time-invariant drifts. During m time-instants, Eq. (8.1) is naturally extended as

$$\hat{x}_i^{(k)} + \epsilon_i \approx \sum_{j=1, j\neq i}^{n} \hat{a}_{i,j} \left(\hat{x}_j^{(k)} + \epsilon_j \right) + \hat{a}_{i,0}, \qquad k = 1, 2, \ldots, m, \qquad (8.2)$$

where $\hat{x}_i^{(k)}$ is the measurement value sensed by ith sensor at kth time-instant. In particular, in order to obtain enough information, we assume $m_0, m > n$. For ith sensor, ϵ_i is a time-invariant drift calibration, which is independent of time-instant k. $\hat{a}_{i,j}$ is the model coefficient when all sensors have unknown time-invariant drifts. We vectorize these variables as $\hat{\mathbf{x}} = [\hat{x}_1^{(1)}, \hat{x}_2^{(1)}, \ldots, \ldots, \hat{x}_n^{(m)}]^\top$, $\hat{\mathbf{a}}_i = [\hat{a}_{i,0}, \ldots, \hat{a}_{i,i-1}, \hat{a}_{i,i+1}, \ldots, \hat{a}_{i,n}]^\top \in \mathbb{R}^n$, $\hat{\mathbf{a}} = [\hat{\mathbf{a}}_1^\top, \hat{\mathbf{a}}_2^\top, \ldots, \hat{\mathbf{a}}_n^\top]^\top \in \mathbb{R}^{n^2}$, and $\epsilon = [\epsilon_1, \epsilon_2, \ldots, \epsilon_n]^\top \in \mathbb{R}^n$.

Note that Eq. (8.2) is essential in our proposed sensor spatial correlation model. Furthermore, the model error in Eq. (8.2) is assumed to follow identical independent Gaussian distribution with zero-mean and unknown precision (inverse variance) δ_0. Therefore, the likelihood function $\mathcal{P}(\hat{\mathbf{x}}|\hat{\mathbf{a}}, \epsilon)$ is defined as follows:

$$\mathcal{P}\left(\hat{\mathbf{x}}|\hat{\mathbf{a}}, \epsilon\right) \propto \exp\left(-\frac{\delta_0}{2} \sum_{i=1}^{n} \sum_{k=1}^{m} \left[\hat{x}_i^{(k)} + \epsilon_i - \sum_{j=1, j\neq i}^{n} \hat{a}_{i,j} \left(\hat{x}_j^{(k)} + \epsilon_j \right) - \hat{a}_{i,0} \right]^2 \right).$$

$$(8.3)$$

However, the likelihood function $\mathcal{P}(\hat{\mathbf{x}}|\hat{\mathbf{a}}, \epsilon)$ cannot be directly used to calibrate drifts using *maximum-likelihood-estimation* (MLE) since it has not enough information. Therefore, we need to give two priors in development.

For all sensors, drifts are assumed to follow identical independent Gaussian distribution with zero-mean and unknown precision δ_ϵ as follows:

$$\mathcal{P}(\epsilon) \propto \exp\left(-\frac{\delta_\epsilon}{2} \sum_{i=1}^{n} \epsilon_i^2 \right). \qquad (8.4)$$

In addition, we assume that the model coefficient $\hat{a}_{i,j}$ follows identical independent Gaussian distribution. Intuitively, $\hat{a}_{i,j}$ has high dependency on $a_{i,j}$ in statistics. Furthermore, the probability density function of $\hat{a}_{i,j}$ is assumed to take a maximum value at $a_{i,j}$. Therefore, the prior mean of $\hat{a}_{i,j}$ is $a_{i,j}$. In addition, in order that each model coefficient $\hat{a}_{i,j}$ is provided with a relatively equal probability to deviate from the corresponding drift-free model coefficient $a_{i,j}$, the precision of model coefficient $\hat{a}_{i,j}$ is defined to be $\lambda a_{i,j}^{-2}$, where λ is a nonnegative hyper-parameter to control the precision of $\hat{a}_{i,j}$. Therefore, each model coefficient $\hat{a}_{i,j}$ follows identical independent Gaussian distribution with $a_{i,j}$ mean and $\lambda a_{i,j}^{-2}$ precision [16–18]. For all model coefficients, we have

$$\mathcal{P}(\hat{\mathbf{a}}) \propto \exp\left(-\sum_{i=1}^{n}\sum_{j=0, j\neq i}^{n} \frac{\lambda}{2a_{i,j}^2}\left(\hat{a}_{i,j} - a_{i,j}\right)^2\right).$$ (8.5)

In order to calibrate drifts for all sensors, the posterior $\mathcal{P}(\hat{\mathbf{a}}, \epsilon|\hat{\mathbf{x}})$ needs to be maximized in MAP estimation manner. According to Bayes' rule, the posterior $\mathcal{P}(\hat{\mathbf{a}}, \epsilon|\hat{\mathbf{x}})$ can be expressed by two priors and the likelihood function as follows:

$$\mathcal{P}(\hat{\mathbf{a}}, \epsilon|\hat{\mathbf{x}}) \propto \mathcal{P}(\hat{\mathbf{x}}|\hat{\mathbf{a}}, \epsilon) \cdot \mathcal{P}(\hat{\mathbf{a}}) \cdot \mathcal{P}(\epsilon).$$ (8.6)

Taking the logarithm, MAP can be transferred to the equivalent formulation as follows:

$$\min_{\hat{\mathbf{a}}, \epsilon} \quad \delta_0 \sum_{i=1}^{n}\sum_{k=1}^{m}\left[\hat{x}_i^{(k)} + \epsilon_i - \sum_{j=1, j\neq i}^{n}\hat{a}_{i,j}\left(\hat{x}_j^{(k)} + \epsilon_j\right) - \hat{a}_{i,0}\right]^2$$

$$+ \lambda \sum_{i=1}^{n}\sum_{j=0, j\neq i}^{n}\frac{1}{a_{i,j}^2}\left(\hat{a}_{i,j} - a_{i,j}\right)^2 + \delta_\epsilon \sum_{i=1}^{n}\epsilon_i^2.$$ (8.7)

There are two challenges for Formulation (8.7): how to handle Formulation (8.7) and how to induce hyper-parameters λ, δ_0, and δ_ϵ.

8.4   Alternating-Based Optimization

Formulation (8.7) is a non-convex problem; thus, it is difficult to obtain an optimal solution. In this section, we propose a fast and efficient alternating-based optimization methodology to handle Formulation (8.7) by alternatively updating in each iteration.

According to the alternating-based methodology, at each iteration, the values of $\hat{\mathbf{a}}$ and ϵ are updated by optimizing Formulation (8.7) w.r.t. $\hat{\mathbf{a}}$ and ϵ. Furthermore, note that with fixed drift calibration variable ϵ, Formulation (8.7) w.r.t. $\hat{\mathbf{a}}$ is regarded as a convex unconstrained quadratic programming (QP) problem. In addition, Formulation (8.7) w.r.t. $\hat{\mathbf{a}}$ can be decomposed into n independent sub-formulations w.r.t. $\hat{\mathbf{a}}_i$ as follows:

$$\min_{\hat{\mathbf{a}}_i} \quad \delta_0 \sum_{k=1}^{m}\left[\hat{x}_i^{(k)} + \epsilon_i - \sum_{j=1, j\neq i}^{n}\hat{a}_{i,j}\left(\hat{x}_j^{(k)} + \epsilon_j\right) - \hat{a}_{i,0}\right]^2$$

$$+ \lambda \sum_{j=0, j\neq i}^{n}\frac{1}{a_{i,j}^2}\left(\hat{a}_{i,j} - a_{i,j}\right)^2,$$ (8.8)

with the first-order optimality condition:

$$\delta_0 \sum_{k=1}^{m} \left(\hat{x}_t^{(k)} + \epsilon_t \right) \left[\sum_{j=1}^{n} \hat{a}_{i,j} \left(\hat{x}_j^{(k)} + \epsilon_j \right) + \hat{a}_{i,0} \right] + \lambda \frac{\left(\hat{a}_{i,t} - a_{i,t} \right)}{a_{i,t}^2} = 0, \qquad (8.9)$$

where $t = 0, 1, \ldots, i - 1, i + 1, \ldots, n$. In particular, we define $\hat{a}_{i,i} \triangleq -1$ and $\hat{x}_0^{(k)} + \epsilon_0 \triangleq 1$. The system of linear equations (8.9) can be handled by Gaussian elimination [19].

In the same manner, with fixed model coefficients $\hat{\mathbf{a}}$, Formulation (8.7) w.r.t. the drift calibration ϵ can also be regarded to be a convex unconstrained QP problem as follows:

$$\min_{\epsilon} \quad \delta_0 \sum_{i=1}^{n} \sum_{k=1}^{m} \left[\hat{x}_i^{(k)} + \epsilon_i - \sum_{j=1, j \neq i}^{n} \hat{a}_{i,j} \left(\hat{x}_j^{(k)} + \epsilon_j \right) - \hat{a}_{i,0} \right]^2 + \delta_\epsilon \sum_{i=1}^{n} \epsilon_i^2, \tag{8.10}$$

with the corresponding first-order optimality condition:

$$\delta_0 \sum_{i=1}^{n} \sum_{k=1}^{m} \left[\hat{a}_{i,t} \left(\sum_{j=1}^{n} \hat{a}_{i,j} \left(\hat{x}_j^{(k)} + \epsilon_j \right) + \hat{a}_{i,0} \right) \right] + \delta_\epsilon \epsilon_t = 0, \qquad (8.11)$$

where $t = 1, 2, \ldots, n$.

Algorithm 1 Alternating-based method

Input: Sensor measurements $\hat{\mathbf{x}}$, prior $\mathbf{a}$ and hyper-parameters λ, δ_0, δ_ϵ.
1: Initialize $\hat{\mathbf{a}} \leftarrow \mathbf{a}$ and $\epsilon \leftarrow \mathbf{0}$;
2: **repeat**
3:     **for** $i \leftarrow 1$ to n **do**
4:         Fix ϵ, solve the system of linear equations (8.9) using Gaussian elimination to update $\hat{\mathbf{a}}_i$;
5:     **end for**
6:     Fix $\hat{\mathbf{a}}$, solve the system of linear equations (8.11) using Gaussian elimination to update ϵ;
7: **until** Convergence
8:
9: **return** $\hat{\mathbf{a}}$ and ϵ.

A local optimum can be obtained by the proposed alternating-based method while the convergence speed and solution quality depend on the initialization of variables. In our proposed framework, two priors are given for model coefficients $\hat{\mathbf{a}}$ and drift calibration ϵ. Therefore, in order to achieve a better convergence speed and solution quality, the prior means $\mathbf{a}$ and $\mathbf{0}$ are used to initialize variables $\hat{\mathbf{a}}$ and ϵ. We continue to update $\hat{\mathbf{a}}$ and ϵ until convergence. The convergence condition is that the relative difference of drift calibration ϵ between current and previous iterations is less than a threshold. In summary, our proposed alternating-based method is shown in Algorithm 1.

8.5 Estimation of Hyper-Parameters

It is important to induce the aforementioned three hyper-parameters so that drifts can be accurately calibrated and meanwhile the over-fitting can be avoided. In this section, cross-validation and EM with Gibbs sampling are used to induce hyper-parameters, respectively.

8.5.1 Unsupervised Cross-Validation

Cross-validation is a simple method to select hyper-parameters. Although there are three hyper-parameters λ, δ_0, δ_ϵ in Formulation (8.7), only two ratios λ/δ_0 and δ_ϵ/δ_0 need to be determined instead of individual hyper-parameters by cross-validation. We partition temperature measurements during m time-instants into s non-overlapping parts. Given each combination of ratios candidates λ/δ_0 and δ_ϵ/δ_0, in each run, one of the s parts is used to estimate the model error and all other $s-1$ parts are used to calculate model coefficients and drift calibration. In the same manner, each run gives a model error e_r $(r = 1, 2, \ldots, s)$ estimated from a part of temperature measurements. The final model error is computed as the average $\bar{e} = (e_1 + e_2 + \cdots + e_s)/s$. Then two ratios λ/δ_0 and δ_ϵ/δ_0 corresponding to the minimum average model error are chosen.

Note that unlike conventional cross-validation [1, 2, 14, 16–18], not any golden value of drift calibration is used in metrics to choose hyper-parameters in model fitting stage. Therefore, in our proposed framework, cross-validation is adopted in an unsupervised-learning-like fashion.

Cross-validation is time-consuming since Algorithm 1 has to be performed for multiple times. Thus, we propose a fast and efficient EM algorithm to determine hyper-parameters in statistical model.

8.5.2 Monte Carlo Expectation Maximization

In this section, MLE is used to determine individual hyper-parameters δ_0, λ, and δ_ϵ. MLE of hyper-parameters is formulated as follows:

$$\max_{\delta_\epsilon, \delta_0, \lambda} \quad \mathcal{P}(\hat{\mathbf{x}}; \delta_0, \lambda, \delta_\epsilon). \tag{8.12}$$

However, the likelihood function $\mathcal{P}(\hat{\mathbf{x}}; \delta_0, \lambda, \delta_\epsilon)$ is intractable. EM algorithm is leveraged to efficiently find a solution to Formulation (8.12). According to EM algorithm, Formulation (8.12) can be taken the logarithm and transferred to be its auxiliary lower bound function [20]. Then, the auxiliary lower bound function is optimized by **E-step** and **M-step** iteratively after the term independent of hyper-parameters is omitted. The detailed derivation can be found in [21]. For convenience, all hyper-parameters are collected as a set Ω.

8.5.2.1 Expectation Step with Gibbs Sampling

In **E-step**, the auxiliary lower bound function can be simplified to be a quantity defined as follows:

$$Q\left(\Omega | \Omega^{\text{old}}\right) = \int \int \mathcal{P}\left(\hat{\mathbf{a}}, \epsilon | \hat{\mathbf{x}}; \Omega^{\text{old}}\right) \ln \mathcal{P}(\hat{\mathbf{x}}, \hat{\mathbf{a}}, \epsilon; \Omega) d\hat{\mathbf{a}} d\epsilon, \qquad (8.13)$$

where Ω^{old} denotes estimated hyper-parameters in the previous iteration.

However, the posterior $\mathcal{P}(\hat{\mathbf{a}}, \epsilon | \mathbf{x}; \Omega^{\text{old}})$ is intractable. There are two main methods to approximate the posterior $\mathcal{P}(\hat{\mathbf{a}}, \epsilon | \mathbf{x}; \Omega)$: variational inference and *Markov chain Monte Carlo* (MCMC). Compared with variational inference, MCMC has the advantage of being non-parametric and asymptotically exact [22]. Therefore, Monte Carlo method is utilized to approximate the quantity as follows:

$$Q\left(\Omega | \Omega^{\text{old}}\right) \approx \frac{1}{L} \sum_{l=1}^{L} \ln \mathcal{P}\left(\hat{\mathbf{x}}, \hat{\mathbf{a}}^{(l)}, \epsilon^{(l)}; \Omega\right), \qquad (8.14)$$

where samples $\hat{\mathbf{a}}^{(l)}$ and $\epsilon^{(l)}$ are obtained from the distribution $\mathcal{P}(\hat{\mathbf{a}}, \epsilon | \hat{\mathbf{x}}; \Omega^{\text{old}})$. L is total amount of samples. In MCMC, there are two main algorithms to obtain samples from the desired distribution $\mathcal{P}(\hat{\mathbf{a}}, \epsilon | \hat{\mathbf{x}}; \Omega^{\text{old}})$: Metropolis Hastings algorithm and Gibbs sampling. Since the rejection rate will be high in complex problems, Metropolis Hastings algorithm has very slow convergence [21]. Therefore, Gibbs sampling is used to obtain samples $\hat{\mathbf{a}}^{(l)}$ and $\epsilon^{(l)}$.

Gibbs sampling has the behavior that one or batch variables are cyclically and repeatedly updated in some particular order at random from conditional distribution. Sampling order is arranged to be $\hat{a}_{1,0}^{(l)}, \ldots, \hat{a}_{1,n}^{(l)}, \hat{a}_{2,0}^{(l)}, \ldots, \hat{a}_{n,n-1}^{(l)}, \epsilon_1^{(l)}, \ldots, \epsilon_n^{(l)}$. In Gibbs sampling, one of key points is derivation of the conditional distribution for each variable. Note that according to Formulation (8.7), the log conditional distribution *w.r.t.* individual variable is quadratic. Therefore, the conditional distribution of each variable is Gaussian distribution as follows:

$$\hat{a}_{p,q} \sim \mathcal{P}\left(\hat{a}_{p,q} | \epsilon, \hat{\mathbf{a}}_{/\hat{a}_{p,q}}, \hat{\mathbf{x}}; \delta_\epsilon, \delta, \lambda\right) = \mathcal{N}\left(\mu_{\hat{a}_{p,q}}, \sigma_{\hat{a}_{p,q}}^{-1}\right),$$

$$\epsilon_t \sim \mathcal{P}\left(\epsilon_t | \epsilon_{/\epsilon_t}, \hat{\mathbf{a}}, \hat{\mathbf{x}}; \delta_\epsilon, \delta, \lambda\right) = \mathcal{N}\left(\mu_{\epsilon_t}, \sigma_{\epsilon_t}^{-1}\right),$$

$$\qquad (8.15)$$

in agreement with (8.4) and (8.5). μ is mean and σ is precision. $\hat{\mathbf{a}}_{/\hat{a}_{p,q}}$ and $\epsilon_{/\epsilon_t}$ denote $\hat{\mathbf{a}}$ but with $\hat{a}_{p,q}$ omitted and ϵ but with ϵ_t omitted.

Before Gibbs sampling, in order to converge to the desired posterior, the warm-start has to be performed if there is no reasonable initialization for samples. Furthermore, it is very hard to judge whether the warm-start is enough [21]. In order to waive the warm-start, a reasonable initialization for samples is adopted in Gibbs sampling. Note that Gibbs sampling is used to obtain samples from the desired posterior $\mathcal{P}(\hat{\mathbf{a}}, \epsilon | \hat{\mathbf{x}}; \Omega^{\text{old}})$ (8.6). As we discussed in Sect. 8.2, Formulation (8.7) is equivalent to MAP estimation of $\hat{\mathbf{a}}$ and ϵ. Thus given hyper-parameters Ω^{old} and measurement values $\hat{\mathbf{x}}$, Gibbs sampling can be initialized by handling Formulation (8.7) to obtain initial samples $\hat{\mathbf{a}}^{(0)}$ and $\epsilon^{(0)}$ satisfying the distribution $\mathcal{P}(\hat{\mathbf{a}}, \epsilon | \hat{\mathbf{x}}; \Omega^{\text{old}})$. As a result, the warm-start can be totally waived.

8.5.2.2 Maximization Step

After L samples are obtained by Gibbs sampling, in **M-step**, we will maximize the approximated quantity as follows:

$$\max_{\Omega} \ \frac{1}{L} \sum_{l=1}^{L} \ln \mathcal{P}\left(\hat{\mathbf{x}}, \hat{\mathbf{a}}^{(l)}, \epsilon^{(l)}; \Omega\right). \tag{8.16}$$

With the first-order optimality condition, that is $dQ/d\Omega = 0$, hyper-parameters λ, δ_0, δ_ϵ can be updated as follows:

$$\lambda = \frac{n^2 L}{\sum_{i=1}^{n} \sum_{j=0, j \neq i}^{n} \sum_{l=1}^{L} \frac{\left(\hat{a}_{i,j}^{(l)} - a_{i,j}\right)^2}{a_{i,j}^2}}, \tag{8.17}$$

$$\delta_0 = \frac{Lmn}{\sum_{l=1}^{L} \sum_{i=1}^{n} \sum_{k=1}^{m} \left[\sum_{j=1}^{n} \hat{a}_{i,j}^{(l)} \left(\hat{x}_j^{(k)} + \epsilon_j^{(l)}\right) + \hat{a}_{i,0}^{(l)}\right]^2}, \tag{8.18}$$

$$\delta_\epsilon = \frac{nL}{\sum_{l=1}^{L} \sum_{i-1}^{n} \epsilon_i^{(l)2}}. \tag{8.19}$$

Here, $\hat{a}_{i,i}^{(l)} \triangleq -1$ and $\hat{x}_0^{(k)} + \epsilon_0^{(l)} \triangleq 1$. We continue to alternate between E-step and M-step until convergence. The convergence condition is that the relative difference of three hyper-parameters between current and previous iterations is less than a threshold. Then hyper-parameters λ, δ, δ_ϵ can be determined.

For convenience, all variables are collected as a set $\Psi = \{\psi_1, \psi_2, \ldots, \psi_{n^2+n}\} = \{\hat{a}_{1,0}, \ldots, \hat{a}_{1,n}, \ldots, \hat{a}_{n,n-1}, \epsilon_1, \ldots, \epsilon_n\}$. In summary, our proposed EM with Gibbs sampling is shown in Algorithm 2.

Algorithm 2 EM with Gibbs sampling

Input: Sensor measurements $\hat{\mathbf{x}}$, prior $\mathbf{a}$;
1: Initialize hyper-parameters Ω;
2: **repeat**
3:     Initialize samples $\Psi^{(0)}$ by Algorithm 1;
4:     **for** $l \leftarrow 1$ to L **do**
5:         **for** $i \leftarrow 1$ to $n^2 + n$ **do**
6:             Sample $\psi_i^{(l)}$ from the desired conditional distribution $\mathcal{N}(\mu_{\psi_i}, \sigma_{\psi_i})$ (8.15) with $\psi_1^{(l)}, \ldots, \psi_{i-1}^{(l)}, \psi_{i+1}^{(l-1)}, \ldots, \psi_{n^2+n}^{(l-1)}$;
7:         **end for**
8:     **end for**
9:     Update hyper-parameters Ω by Equations (8.17), (8.18) and (8.19);
10: **until** Convergence
11:
12: **return** hyper-parameters Ω.

8.6 Experimental Results

The in-building temperature data are used to test our proposed framework. We use several sensors to calibrate drifts. All data is directly generated from EnergyPlus as shown in Fig. 8.5. As shown in Fig. 8.6, two building benchmarks, Hall [23] with Washington, D.C weather and Secondary School [24] with Chicago weather, are simulated by EnergyPlus to generate the ground-truth in-building temperatures, which are used to test our proposed framework. The temperature sampling period is set to be 1 h.

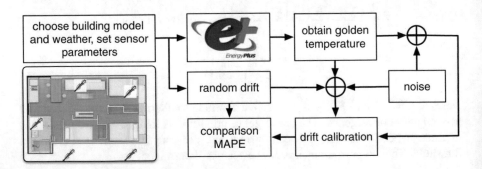

Fig. 8.5 The generated simulation data

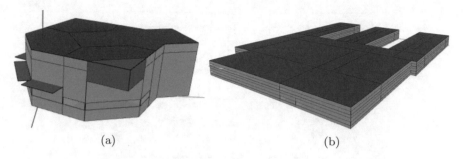

Fig. 8.6 Benchmark: (**a**) Hall; (**b**) Secondary School

In practice, both drift and measurement noise need to be carefully considered and reasonably set to close to real temperature measurement. Because of a slow-aging effect, time effects on sensor performance are not considered in our experiments. Drift is set to be time-invariant while measurement noise is set to be time-variant. According to the sensors' performance shown in Fig. 8.2a, two low-cost temperature sensors, MCP9509 with accuracy ± 4.5 °C and LM335A with accuracy ± 5 °C as shown in Fig. 8.2b,c, are chosen to set drift variance, respectively. According to the triple standard deviation, we set two drift variances to be $\sigma^2 = (4.5/3)^2 = 2.25$ and $\sigma^2 = (5/3)^2 = 2.78$. In addition, according to our survey, the noise variance is set to be 0.001. All temperature measurements are generated by adding noise.

The time-instant number needs to be reasonably set to meet practical application and accurately calibrate sensor drifts. We assume the temperature measurements are drift-free during first $m_0 = 240$ time-instants (first 10 days). And during $m = 60$ time-instants (60 h), the temperature measurements with drifts are used to test our proposed framework.

TSBL [11] and the proposed framework with cross-validation and EM are used to calibrate sensor drifts, respectively. All methods are implemented by `Python 2.7` on 12-core Linux machine with 256 G RAM and 2.80 GHz. 100 combinations of hyper-parameters ratios and $s = 5$ folds are set in cross-validation. Since the warm-start is waived in Gibbs sampling, in order to achieve a better trade-off between accuracy and runtime, only $L = 10$ samples are generated to perform Monte Carlo approximation (8.14), and three hyper-parameters λ, δ_0, δ_ϵ are initialized to be 10^3, 10^{-4}, and 10^{-3} in EM. The convergence criterion thresholds are set to be 10^{-8} and 10^{-2} in Algorithms 1 and 2.

As mentioned in Sect. 8.2, the drift calibration accuracy is evaluated by using MAPE defined as follows:

$$\text{MAPE} = \frac{1}{nm} \sum_{k=1}^{m} \sum_{i=1}^{n} \left| \frac{\hat{\epsilon}_i^{(k)} - \epsilon_i}{\epsilon_i} \right|, \tag{8.20}$$

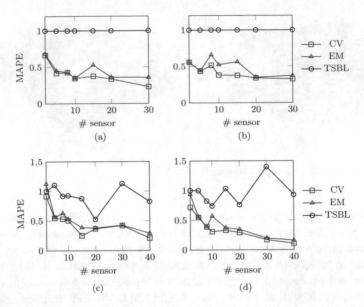

Fig. 8.7 Drift variance is set to (**a,c**) 2.25; (**b,d**) 2.78; Benchmark: (**a,b**) Hall; (**c,d**) Secondary school

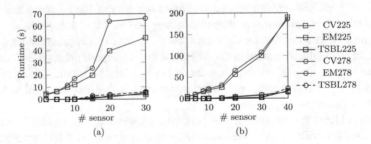

Fig. 8.8 Runtime vs. # sensor on (**a**) Hall; (**b**) Secondary school

where $\hat{\epsilon}_i^{(k)}$ is the estimated calibration. Specifically, in our proposed framework, $\hat{\epsilon}_i^{(k)} = \hat{\epsilon}_i$. The sensor drift calibration performances of accuracy and runtime are shown in Figs 8.7 and 8.8.

As shown in Fig. 8.8, TSBL has acceptable computational overhead even if its computational complexity is dominated by multiple matrix inversion operations. However, as shown in Fig. 8.7, TSBL has the worst performance and robust for drifts calibration. In fact, temperature signals lie in time-variant subspace since in-building temperatures are influenced by multiple time-variant factors, e.g., weather. As a result, TSBL cannot achieve an obvious drift calibration.

Unlike TSBL, the proposed spatial correlation model can calibrate drifts even if temperature signals lie in time-variant subspace. Therefore, as shown in Fig. 8.7, the proposed framework with either cross-validation or EM outperforms TSBL in accuracy. Besides, the proposed drift calibration framework with cross-validation can achieve the best accuracy. However, as shown in Fig. 8.8, cross-validation has heavy computational overhead since we need to run Algorithm 1 for multiple times. Compared with cross-validation and TSBL, EM with Gibbs sampling has lower computation complexity since less samples are generated to perform Monte Carlo approximation and EM can achieve fast convergence. However, as shown in Fig. 8.7, the proposed framework with EM cannot achieve the best accuracy since EM with Gibbs sampling is an approximation method.

As shown in Fig. 8.7, because of incremental correlation, the more sensors can achieve the more accuracy of drift calibration by using our proposed framework. In practice, when less sensors need to be calibrated, in order to achieve a better accuracy, cross-validation can be used to determine hyper-parameters within a reasonable response time, e.g., 1 min. While more sensors need to be calibrated, EM with Gibbs sampling can be used to determine hyper-parameters so that sensor measurement accuracy can be improved to a tolerable level within acceptable runtime. The proposed calibration framework with EM can achieve robust drift calibration and a better trade-off between accuracy and runtime.

8.7 Conclusion

In this paper, a sensor spatial correlation model has been proposed to perform drift calibration. Thanks to spatial correlation, the unknown actual temperature measured by each sensor is linearly expressed by all other sensors. The priors for model coefficients and drift calibration are applied to MAP estimation. MAP estimation is then formulated as a non-convex problem with three hyper-parameters, which is handled by the proposed alternating-based method. Cross-validation and EM with Gibbs sampling are used to determine hyper-parameters, respectively. Experimental results show that on benchmarks simulated from EnergyPlus, the proposed framework with EM can achieve a robust drift calibration and better trade-off between accuracy and runtime.

References

1. X. Chen, X. Li, S.X.-D. Tan, Overview of cyber-physical temperature estimation in smart buildings: from modeling to measurements, in *INFOCOM Workshops* (2016), pp. 251–256
2. B. Lin, B. Yu, Smart building uncertainty analysis via adaptive lasso. IET Cyber-Phys. Syst. Theory Appl. **2**(1), 42–48 (2017)
3. Q. Zhu, A. Sangiovanni-Vincentelli, S. Hu, X. Li, Design automation for cyber-physical systems. Proc. IEEE **106**(9), 1479–1483 (2018)
4. K. Ni, N. Ramanathan, M.N.H. Chehade, L. Balzano, S. Nair, S. Zahedi, E. Kohler, G. Pottie, M. Hansen, M. Srivastava, Sensor network data fault types. ACM Trans. Sens. Netw. (TOSN) **5**(3), 25 (2009)

5. *Engineer's Guide to Accurate Sensor Measurements* (2016). http://download.ni.com/evaluation/daq/25188_Sensor_WhitePaper_IA.pdf
6. *Findchips* (2018). https://www.findchips.com
7. Y. Wang, A. Yang, Z. Li, P. Wang, H. Yang, Blind drift calibration of sensor networks using signal space projection and Kalman filter, in *IEEE 10th International Conference on Intelligent Sensors, Sensor Networks and Information Processing (ISSNIP)* (2015), pp. 1–6
8. M. Takruri, S. Challa, R. Yunis, Data fusion techniques for auto calibration in wireless sensor networks, in *International Conference on Information Fusion* (2009), pp. 132–139
9. M. Takruri, S. Rajasegarar, S. Challa, C. Leckie, Online drift correction in wireless sensor networks using spatio-temporal modeling, in *International Conference on Information Fusion* (2008), pp. 1–8
10. L. Balzano, R. Nowak, Blind calibration of sensor networks, in *International Conference on Information Processing in Sensor Networks (IPSN)* (2007), pp. 79–88
11. Y. Wang, A. Yang, Z. Li, X. Chen, P. Wang, H. Yang, Blind drift calibration of sensor networks using sparse Bayesian learning. IEEE Sensors J. **16**(16), 6249–6260 (2016)
12. Z. Zhang, B.D. Rao, Sparse signal recovery with temporally correlated source vectors using sparse Bayesian learning. IEEE J. Sel. Top. Sign. Proces. **5**(5), 912–926 (2011)
13. S. Ling, T. Strohmer, Self-calibration and bilinear inverse problems via linear least squares. SIAM J. Imag. Sci. (SIIMS) **11**(1), 252–292 (2018)
14. X. Chen, X. Li, S. X.-D. Tan, From robust chip to smart building: CAD algorithms and methodologies for uncertainty analysis of building performance, in *IEEE/ACM International Conference on Computer-Aided Design (ICCAD)* (2015), pp. 457–464
15. *2-Terminal IC Temperature Transducer* (2013). https://www.analog.com/media/en/technical-documentation/data-sheets/AD590.pdf
16. F. Wang, P. Cachecho, W. Zhang, S. Sun, X. Li, R. Kanj, C. Gu, Bayesian model fusion: large-scale performance modeling of analog and mixed-signal circuits by reusing early-stage data. IEEE Trans. Comput. Aided Des. Integr. Circuits Syst. (TCAD) **35**(8), 1255–1268 (2016)
17. Q. Huang, C. Fang, F. Yang, X. Zeng, X. Li, Efficient multivariate moment estimation via Bayesian model fusion for analog and mixed-signal circuits, in *ACM/IEEE Design Automation Conference (DAC)* (2015), p. 169
18. Q. Huang, C. Fang, F. Yang, X. Zeng, D. Zhou, X. Li, Efficient performance modeling via dual-prior Bayesian model fusion for analog and mixed-signal circuits, in *ACM/IEEE Design Automation Conference (DAC)* (2016), pp. 1–6
19. G.H. Golub, C.F. Van Loan, *Matrix Computations* (JHU Press, Baltimore, 2012)
20. K. Ganchev, B. Taskar, J. Gama, Expectation maximization and posterior constraints, in *Conference on Neural Information Processing Systems (NIPS)* (2008), pp. 569–576
21. C. Robert, *Machine Learning, a Probabilistic Perspective* (Taylor & Francis, London, 2014)
22. T. Salimans, D. Kingma, M. Welling, Markov chain Monte Carlo and variational inference: bridging the gap, in *International Conference on Machine Learning (ICML)* (2015), pp. 1218–1226
23. *OpenStudio®* (2018). https://www.openstudio.net
24. *National Renewable Energy Laboratory OpenStudio Standards* (2018). https://github.com/NREL/openstudio-standards

Chapter 9
Fault Tolerance-Aware Design Technique for Cyber-Physical Digital Microfluidic Biochips

Xiaodao Chen, Yuewei Wang, Chaowei Wan, and Xiaohui Huang

9.1 Introduction

The digital microfluidic biochip (DMFB), considered as a lab-on-a-chip, has been widely applied in Biomedicine and Biotechnology. It is a scalable system with liquid droplets which can be controlled by electrowetting-on-dielectric (EWOD) [1]. As shown in Fig. 9.1, the droplets can be driven in a 2-D electrodes array where detectors are used to check the operations. Bio-reactions can be achieved by mixing droplets. To implement bio-assays, each on-chip operation needs to be scheduled and placed on its corresponding electrode on the DMFB. Based on this, the DNA analysis, the drug automation, and clinical diagnostics can be achieved [2]. Comparing to the traditional laboratory procedures, DMFBs outperform in terms of accuracy, cost efficiency, and time cost [3]. The reason is that DMFBs have advantages in sample holding, reagent manipulation, droplets detection, etc. [4, 5]. Meanwhile, along with the usages of DMFB spreading, the complexity of DMFBs is also increased. Thus, the computer aided design (CAD) methodologies are embraced with DMFB design for the effectiveness and efficiency [6, 7].

In the past decades, CAD methodologies for digital microfluidic biochips are mainly focused on the offline physical design and synthesis, such as literatures [8–10]. The CAD design for digital microfluidic biochips has been developed significantly, but there are still some challenges which the DMFBs design has to face. The major one of them is the reliability issue during on-chip reactions taking place [11]. Since on-chip reactions can be delicate and sensitive, any flaws can lead to incorrect results of bio-reactions. The flaws can be caused by the chip manufacture or the liquid droplet variation. To overcome this issue, online recover scheme has been developed. Once there is an error detected on DMFBs, this recover

X. Chen (✉) · Y. Wang · C. Wan · X. Huang
School of Computer Science, China University of Geosciences, Wuhan, China

© Springer Nature Switzerland AG 2020
S. Hu, B. Yu (eds.), *Big Data Analytics for Cyber-Physical Systems*,
https://doi.org/10.1007/978-3-030-43494-6_9

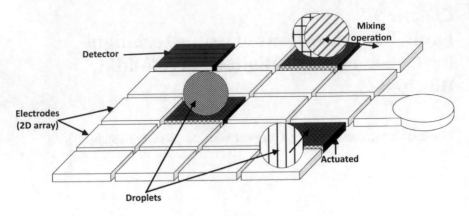

Fig. 9.1 Illustration of a digital microfluidic biochip

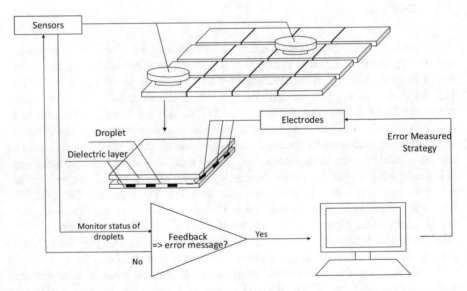

Fig. 9.2 CPS flow for fault tolerance-aware biochip design

scheme can rollback the on-chip processing to the previous bio-reaction and re-implement the procedure which encounter the error [12, 13]. The error detection procedure can be handled by error detecting sensors which called checkpoints [1]. Thus, the robustness of biochips can be achieved.

Figure 9.2 illustrates how the CPS technique is adopted in the processing of DMFBs. On a cyber-physical DMFB, a feedback controller and two sensors which are checkpoints are employed. During the processing of bio-reactions on this DMFB, checkpoints monitor the status of droplets when it moves through it.

The monitored feedback results are sent back to the feedback controller in real-time. These feedback results can guide the controller determining whether the bio-reactions are executed correctly and whether a bio-reaction needs rollback. Once an error message is detected by the controller, the error measured strategies, which rollback bio-reactions, are invoked to be executed by the help of electrodes.

Since the on-chip resource is limited for the DMFB design, where to place the checkpoints for the fault tolerance-aware biochip design becomes critical. Solutions for minimizing checkpoint with full coverage of on-chip bio-reactions are highly desired. In this chapter, an integer linear programming (ILP) technique-based checkpoints placement approach is proposed for cyber-physical DMFB design.

The rest of this chapter is organized as follows: Sect. 9.2 presents the problem formulation, Sect. 9.3 describes the proposed algorithm, Sect. 9.4 presents experimental results, and Sect. 9.5 concludes the whole chapter.

9.2   Problem Formulation

Due to the fault caused by randomness during the DMFB manufacture and on-chip reaction, on-chip checkpoints are utilized in the DMFB system to detect errors during the droplet passing through it during the procedure of biomedical reactions [9]. When droplets are detected as there is error with it in terms of volume or concentration [14], the related bio-reactions be re-implemented.

Suppose that there is a set of bio-reactions and each bio-reaction route on-chip droplets from their source to their destination. Given an on-chip routing solution, the **Objective** of our proposed methods is to minimize the number of checkpoints. The **Input** of minimization problem consists of a set of bio-reactions, and the routing paths of each bio-reaction. The **Constraints** of the minimization problem include the following aspects.

- Each droplet needs to move through at least one checkpoint before it reaches its destination.
- checkpoints cannot be placed at cells which need to be reserved for some other design purposes.
- For each cell, there can be placed at most one checkpoint.

9.3   The Algorithm

The integer linear programming (ILP) technique is utilized to solve checkpoints placement problem which has been discussed in Sect. 9.2.

Let a and b denote the size of a DMFB, then there are $a \times b$ cells on the DMFB, which can be represented as Γ, a set of available cells for checkpoint placement. Note that, due to the hardware architecture constraints, the maximum number of checkpoint locating in a cell is limited to not greater than one. Each checkpoint located at the cell must be an integer and in this chapter C_* is used to denote it. For example, C_i represents the case that whether a checkpoint is placed at the cell i. If a checkpoint is placed at the cell i, the value of C_i is equal to 1. Otherwise, the value of C_i is equal to 0. As shown in Fig. 9.3, a checkpoint can be placed in a cell but two checkpoints cannot be placed in a cell. The constraint of the ILP formulation for the parameter C_i can be described as Eq. (9.1).

$$0 \le C_i \le 1, C_i \in N, \forall i \in \Gamma, \tag{9.1}$$

where N denotes the set of integer numbers. Theoretically, all cells on DMFB are able to accommodate a checkpoint. In practice, several cells are reserved for other design purpose, and these cells cannot be placed by checkpoints. For the illustration convenience, these cells are called reserved cells. As shown in Fig. 9.4, checkpoints cannot be placed at reserved cells. To comprehensively consider this situation, according to the definition of parameter C_*, the constraint can be illustrated as Eq. (9.2).

$$C_j = 0, \forall j \in \Lambda, \Lambda \subseteq \Gamma, \tag{9.2}$$

Fig. 9.3 Two or more checkpoints cannot placed in a cell

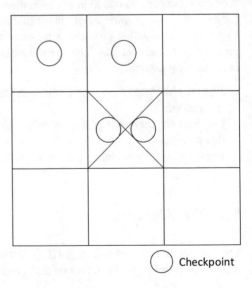

Checkpoint

Fig. 9.4 No checkpoint can be placed at a reserved cell

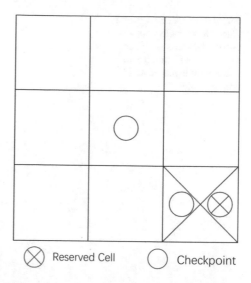

$\bigotimes$ Reserved Cell $\bigcirc$ Checkpoint

where Λ is a set including cells which cannot be equipped by a checkpoint. In addition, the checkpoint is utilized to guarantee the quality of the droplet in the chemical experiment. All droplets routing have to pass through at least one checkpoint, such that the correctness of the experiment can be verified in real-time. Let Φ_τ denote the set of cells on the τth route of the droplet and $C_{\tau k}$ denotes the possess the situation of the kth cell. As shown in Fig. 9.5, one and two checkpoints are located at a route in sub-figure (a) and sub-figure (b), respectively. But there is no checkpoint located at sub-figure (c) and this is not allowed, since there is a routing path which is not monitored. To describe this, the corresponding constraint equations are shown as follows:

$$\sum C_{\tau k} \geq 1, \forall k \in \Phi_\tau \tag{9.3}$$

The object of the checkpoint placement is to minimize the total number of on-chip checkpoints, because $C_* = 1$ is defined to illustrate the cell which contains a checkpoint. Combining constraints illustrated above, the ILP formulation can be described as follows:

$$min \ \sum_i^\Gamma C_i$$
$$s.t.$$
$$0 \leq C_i \leq 1, C_i \in N, \forall i \in \Gamma \tag{9.4}$$
$$C_j = 0, \forall j \in \Lambda, \Lambda \subseteq \Gamma$$
$$\sum C_{\tau k} \geq 1, \forall k \in \Phi_\tau$$

To clearly illustrate the algorithm, a toy example is presented as follows.

Fig. 9.5 (**a**) A checkpoint is placed at a droplet route, (**b**) two checkpoints are placed at a droplet route, and (**c**) no checkpoint is placed at a droplet route

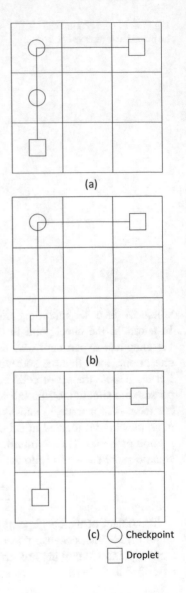

As shown in Fig. 9.6, the DMFB is in 4×4 shape and there are 16 cells in total. Corresponding to each cells, a set illustrated whether they are possessed by a checkpoint is presented as $\{C_1, C_2, C_3, \ldots, C_{16}\}$. Equation (9.1) can be transformed into as follows:

$$0 \le C_1, C_2, C_3, \ldots, C_{16} \le 1, C_1 \in N, C_2 \in N, \ldots, C_{16} \in N \qquad (9.5)$$

Fig. 9.6 A toy example of checkpoint placement

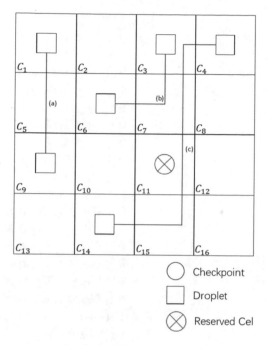

From Fig. 9.6, the cell at third column and third row has a broken mark which stands for this cell is a reserved cell. It is 11th cell of total cells, Eq. (9.2) can be transformed into as follows:

$$C_{11} = 0 \tag{9.6}$$

There are three routes presented in Fig. 9.6 and at least one checkpoint is required to monitor one routing path. For example, the first route (a) involves cells 1, 5, and 9. In these four cells, one or more than one cell need to be placed with checkpoints and their corresponding value C_* need to be equal to 1. Equation (9.3) can be transformed into as follows:

$$
\begin{aligned}
C_1 + C_5 + C_9 &\geq 1 \\
C_3 + C_6 + C_7 &\geq 1 \\
C_3 + C_4 + C_7 + C_{11} + C_{14} + C_{15} &\geq 1
\end{aligned} \tag{9.7}
$$

To minimize the number of checkpoints, the target problem can be formulated as shown in Eq. (9.4).

$$min \ \sum_{i=1}^{16} C_i$$
$$s.t. \ \ 0 \leq C_1 \leq 1, C_1 \in N$$
$$0 \leq C_2 \leq 1, C_2 \in N$$
$$0 \leq C_3 \leq 1, C_3 \in N$$
$$0 \leq C_4 \leq 1, C_4 \in N$$
$$0 \leq C_5 \leq 1, C_5 \in N$$
$$0 \leq C_6 \leq 1, C_6 \in N$$
$$0 \leq C_7 \leq 1, C_7 \in N$$
$$0 \leq C_8 \leq 1, C_8 \in N$$
$$0 \leq C_9 \leq 1, C_9 \in N$$
$$0 \leq C_{10} \leq 1, C_{10} \in N \quad\quad\quad (9.8)$$
$$0 \leq C_{11} \leq 1, C_{11} \in N$$
$$0 \leq C_{12} \leq 1, C_{12} \in N$$
$$0 \leq C_{13} \leq 1, C_{13} \in N$$
$$0 \leq C_{14} \leq 1, C_{14} \in N$$
$$0 \leq C_{15} \leq 1, C_{15} \in N$$
$$0 \leq C_{16} \leq 1, C_{16} \in N$$
$$C_{11} = 0$$
$$C_1 + C_5 + C_9 \geq 1$$
$$C_3 + C_6 + C_7 \geq 1$$
$$C_3 + C_4 + C_7 + C_{11} + C_{14} + C_{15} \geq 1$$

9.4   Experimental Results

Simulation results have been performed to evaluate the proposed approach which targets at the minimizing the total number of checkpoint on a DMFB. The proposed algorithm is implemented in python 3.6. A python package, which is named *lp_solve* 5.5.2.5 (http://lpsolve.sourceforge.net/5.5/), is utilized for solving linear programs. The approach was tested on a workbench with 3.6 GHz Intel $i7 - 7700$ CPU and 16 GB memory. The simulation was conducted on a standard benchmark [15], which has two main bio-reactions. These bio-reactions are the vitro reaction and the protein reaction. We use the algorithm from [16] to generate the placement. Based on it, a heuristic algorithm is used to generate the biochip routing as the input of the proposed algorithm. Details of the input testcase are listed in Table 9.1.

The experimental results without considering reserved cells are shown in Table 9.2. The following observations are made:

- The number of checkpoints obtained by the proposed method is the optimal solution. Vitro 1 has eight checkpoints, vitro 2 has eight checkpoints, vitro 3 has four checkpoints, and protein 3 has five checkpoints.

Table 9.1 Testcase specifications

Testcase	Chip size	# of modules	Routing wire length	Complete time
vitro 1	9×9	64	327	56
vitro 2	8×8	64	370	59
vitro 3	7×7	64	288	56
protein 3	11×11	103	1102	194

Table 9.2 Results without considering reserved cells

Testcase	Number of checkpoints	Runtime (ms)
Vitro 1	8	13.10
Vitro 2	8	13.16
Vitro 3	4	11.75
Protein 3	5	27.77

- The relation between the complexity of the biochip and the number of check-points is not monotonic. Even though the protein 3 has the largest chip size, the longest routing length, and the greatest number of on-chip modules, the number of its checkpoints is not the most significant one among the testcases.
- The chip size can impact the runtime of the proposed approach. The protein 3 has the largest chip size among the testcases and it has the longest running time. The reason is that a larger size of the biochip can introduce more number of variables in the ILP formulation.

In Fig. 9.7, results of vitro 3 have been shown. Figure 9.7a is the initial input which is a biochip design after placement and routing. Viewing from the top, Fig. 9.7a can be represented as Fig. 9.7b. The placement of the checkpoints, which is the result, is shown in Fig. 9.7c. Note that green circles here denote checkpoints. Figure 9.7d is the 3-D version of the results, checkpoints are denoted by green cylinders.

We also tested the proposed algorithm with reserved cells. As shown in Fig. 9.8a, red rectangles denote reserved cells which cannot be occupied by checkpoints. For this initial input, checkpoint placement results are presented in Fig. 9.8b. Comparing Fig. 9.8a and b, the only difference is that Fig. 9.8a has reserved cells. Thus, the number of checkpoints is increased from 4 to 5.

9.5 Conclusion

DMFBs can conduct complex experiments accurately and effectively by utilizing multiple electro wetting-on-dielectric which can control chemical liquid droplets precisely. To guarantee real-time correctness of experiments, cyber-physical system technique is applied to DMFBs for checkpoints placement. These checkpoints can be equipped at cells to monitor the state of electrode cells in real-time, and to facil-

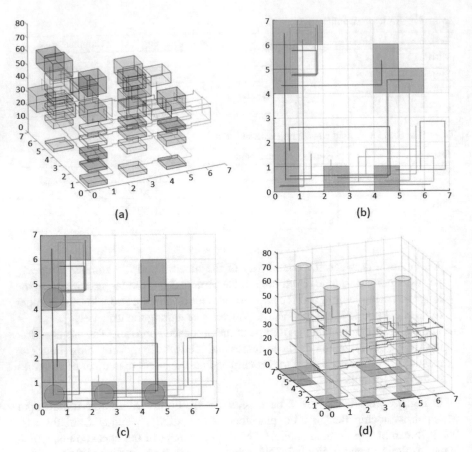

Fig. 9.7 Results on vitro 3 (**a**) Initial input in 3D. (**b**) Initial input viewing from top. (**c**) Placement result viewing from top. (**d**) Placement result in 3D

itate the system recovery from errors. To solve the checkpoints placement problem, an ILP technique-based approach is proposed in this chapter. In experimental parts, "vitro" and "protein" testbenches are utilized under scenarios that both considering cases with and without reserved cells. In both scenarios, the effectiveness and efficiency of the proposed method have been demonstrated.

Acknowledgement This work was supported in part by the National Natural Science Foundation of China (No. 61501411).

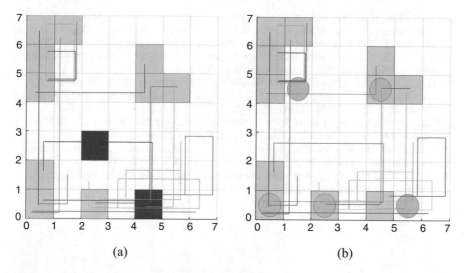

Fig. 9.8 Results on vitro 3 with considering reserved cells. (**a**) Input with reserved cells. (**b**) Placement result for input with reserved cells

References

1. G.-R. Lu, G.-M. Huang, A. Banerjee, B.B. Bhattacharya, T.-Y. Ho, H.-M. Chen, On reliability hardening in cyber-physical digital-microfluidic biochips, in *2017 22nd Asia and South Pacific Design Automation Conference (ASP-DAC), January* (2017)
2. S. Ali, M. Ibrahim, O. Sinanoglu, K. Chakrabarty, R. Karri, Security assessment of cyber-physical digital microfluidic biochips. IEEE/ACM Trans. Comput. Biol. Bioinf. **13**, 445–458 (2016)
3. O. Keszocze, R. Wille, K. Chakrabarty, R. Drechsler, A general and exact routing methodology for digital microfluidic biochips, in *2015 IEEE/ACM International Conference on Computer-Aided Design (ICCAD)* (2015), pp. 874–881
4. N. Li, A. Tourovskaia, A. Folch, Biology on a chip: microfabrication for studying the behavior of cultured cells. Crit. Rev. Biomed. Eng. **31**, 423 (2003)
5. B.H. Weigl, R.L. Bardell, C.R. Cabrera, Lab-on-a-chip for drug development. Adv. Drug Deliv. Rev. **55**(3), 349–377 (2003)
6. Z. Yang, K. Chakrabarty, Cross-contamination avoidance for droplet routing in digital microfluidic biochips, in *Design, Automation and Test in Europe Conference and Exhibition* (2009), pp. 1290–1295
7. K.F. Bohringer, Modeling and controlling parallel tasks in droplet-based microfluidic systems. IEEE Trans. Comput. Aided Des. Integr. Circuits Syst. **25**(2), 334–344 (2006)
8. P.-H. Yuh, S. Sapatnekar, C.-L. Yang, Y.-W. Chang, A progressive-ILP based routing algorithm for cross-referencing biochips, in *45th ACM/IEEE Design Automation Conference, 2008. DAC 2008* (IEEE, Piscataway, 2008), pp. 284–289
9. Y.H. Chen, C.L. Hsu, L.C. Tsai, T.W. Huang, T.Y. Ho, A reliability-oriented placement algorithm for reconfigurable digital microfluidic biochips using 3-d deferred decision making technique. IEEE Trans. Comput. Aided Des. Integr. Circuits Syst. **32**(8), 1151–1162 (2013)
10. A. Grimmer, Q. Wang, H. Yao, T.-Y. Ho, R. Wille, Close-to-optimal placement and routing for continuous-flow microfluidic biochips, in *2017 22nd Asia and South Pacific Design Automation Conference (ASP-DAC)* (IEEE, Piscataway, 2017), pp. 530–535

11. Y.-H. Chen, C.-L. Hsu, L.-C. Tsai, T.-W. Huang, T.-Y. Ho, A reliability-oriented placement algorithm for reconfigurable digital microfluidic biochips using 3-d deferred decision making technique. IEEE Trans. Comput. Aided Des. Integr. Circuits Syst. **32**(8), 1151–1162 (2013)
12. M. Ibrahim, K. Chakrabarty, Efficient error recovery in cyberphysical digital-microfluidic biochips. IEEE Trans. Multi-Scale Comput. Syst. **1**(1), 46–58 (2015)
13. Y. Luo, K. Chakrabarty, T.-Y. Ho, Real-time error recovery in cyberphysical digital-microfluidic biochips using a compact dictionary. IEEE Trans. Comput. Aided Des. Integr. Circuits Syst. **32**(12), 1839–1852 (2013)
14. Y. Luo, K. Chakrabarty, T.Y. Ho, Error recovery in cyberphysical digital microfluidic biochips. IEEE Trans. Comput. Aided Des. Integr. Circuits Syst. **32**(1), 59–72 (2013)
15. C. Liao, S. Hu, Physical-level synthesis for digital lab-on-a-chip considering variation, contamination, and defect. IEEE Trans. Nanobiosci. **13**(1), 3–11 (2014)
16. C. Wan, X. Chen, D. Liu, A multi-objective driven placement technique for digital microfluidic biochips. J. Circ. Syst. Comput. **28**, 1950076 (2019)

Chapter 10
Uncertainty Analysis and Optimization in Cyber-Physical Systems of Reservoir Production

Hui Zhao, Lin Cao, Xingkai Zhang, and Xuewei Ning

10.1 Introduction

The goal of reservoir closed-loop production optimization management [1–4] is maximizing the cumulative oil production or economic benefits by adjusting the well controls (developing strategy) for the expected reservoir life. Based on production data and reservoir numerical simulation technique, closed-loop production optimization management alters history matching and production optimization efficiently, which could update indeterminate geological model and estimates the flow distribution, ultimately, finds out the optimized strategy and realizes the maximized benefits.

History matching (data assimilation), an inverse problem [5], refers to the determination of the plausible physical properties of the system, or information about these properties, given the observed response of the system to some stimulus. The observed response will be referred to as observed data. Note that measured or observed data is different from the problem data introduced in the definition of a well-posed problem. Besides, parameters of points between wells are calculated by interpolation method according to known measured data of wells which causes errors during calculation. Those errors of measurements and calculations cannot be eliminated. But the database will be refreshed and assimilated with previous data to simulate a group of models of which expectation could approximate real underground situation. The novel way based on simulating the expectation of multi-model is called EnKF [6–12]. The reason we put the updated production observed data again into reservoir numerical simulation for data assimilation is to verify and decrease the uncertainty of sensitivity and real geological model [13].

H. Zhao (✉) · L. Cao · X. Zhang · X. Ning
College of Petroleum Engineering, Yangtze University, Wuhan, China
e-mail: zhaohui@yangtzeu.edu.cn

© Springer Nature Switzerland AG 2020
S. Hu, B. Yu (eds.), *Big Data Analytics for Cyber-Physical Systems*,
https://doi.org/10.1007/978-3-030-43494-6_10

Fig. 10.1 Process of
reservoir closed-loop
production optimization
management

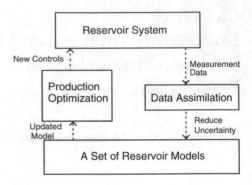

The other key part is production optimization of reservoir, on the contrary, which is a forward problem. In a forward problem, the physical properties of reservoir systems are known, and a deterministic method is available for calculating the response or outcome of the system to a known stimulus. The production optimization of reservoir, on the base of simulated reservoir model to analyze and verify the most possible flood distribution, is aiming at establishing the optimized schedule in future production by reservoir development which is based on history matching and optimization control. Briefly, what we called reservoir closed-loop production optimization management is the process of strategy changing after optimization and then updates or verifies the established reservoir model by adding new production observed data, return to initial step, repeat the same process. The process of history matching and production optimization will not stop until the maximized ultimate oil recovery or economic benefits is output. It can be stated that both history matching and production optimization are typical optimization problems.

Current methods for closed-loop production optimization are categorized by two groups: adjoint gradient algorithm, such as conjugate gradient, quasi-Newton, and derivative-free algorithm, such as SPSA, ensemble-based optimization (EnOpt), new unconstrained optimization algorithm (NEWUOA), simplex derivative-pattern search method (SID-PSM) [1, 7, 9]. However, the gradient of function in reservoir barely could be solved out, especially in optimization. Scholars began working on this problem and did a lot of research, then adopted the adjoint gradient method. Coding and calculating adjoint matrix in program need obtaining gradient. Obviously, solving sensitivity matrix (gradients) is too complicated to achieve the goal. Furthermore, in one step, solving gradient has to be calculated two times, forward and inverse, under the condition of full-implicit in reservoir simulation, which does harm to efficiency. Though the gradient-based method shows fast speed in example application, it still cannot be used in solving real reservoir problems. In addition, because of uncertainty of single-model reservoir, large-scale parameters are adjusted artificially, making more uncertainties. The search direction of SPSA [11, 12, 14] keeps uphill, so that convergence can be guaranteed. In reality, established single-model barely reflects every aspect of real underground.

To decrease the uncertainty and risks of single-model, we take SPSA as effective algorithm based on multi-model reservoir which is called robust optimization [15]. Example application in this article demonstrates a satisfied exploitation effect and supplies theoretical support for practical reservoir closed-loop management [16, 17].

10.2 Data Assimilation

EnKF is the most widespread method in history matching which adopted in Marine dynamics and meteorology the very beginning, but it attracts enough attention after Naevdal [6] and his team led this way into history matching of petroleum engineering in 2002. The basic idea of the EnKF method is designing a set of model realizations by the Monte Carlo method. Each realization of this set will be seen as objective and updates itself by assimilating observed data. Updated average model of set could be the maximum a posteriori of reality, and covariance of set approximates to background error covariance.

By the EnKF method, vector s should first be set in model inversion, which includes geological static and state parameters. Denote s as

$$s = \begin{bmatrix} m \\ y \end{bmatrix} \tag{10.1}$$

In Eq. (10.1), m is a vector of each grid in model, which includes static parameters, such as permeability and porosity; y is another vector of each grid in model, which includes state parameters, such as pressure, saturation.

Suppose that there is an ensemble of N_e model realizations, in the jth model, in the nth simulation step, EnKF data assimilation equation is

$$s_j^{u,n} = s_j^{p,n} + C_{S^{p,n}, D^{p,n}} (C_{D^{p,n}, D^{p,n}} + C_{D^n})^{-1} (d_{uc,j}^n - d_j^{p,n}) \tag{10.2}$$

where $s_j^{p,n}$ is the model realization before the nth step update; $s_j^{u,n}$ is the model realization after the nth step update; $C_{D^n}^n$ is the covariance matrix of observed data in the nth calculation step; $d_j^{p,n}$ is the observed data vector calculated from the jth model; $d_{uc,j}^n$ is the observed data vector relates to the jth model realization which is calculated by observed data vector plus error perturbation vector. There is a theoretical relationship between $d_{uc,j}^n$ and d_{obs}^n which yields

$$d_{uc,j}^n = d_{obs}^n + C_{D^n}^{1/2} Z_d \tag{10.3}$$

where Z_d is a random vector with normal probability distribution; $C_{D^n}^n$ is a square matrix, which is decomposed by the Cholesky [18] method at well as satisfy $C_{D^n}^{1/2} C_{D^n}^{T/2} = C_{D^n}$. Calculation equations of sensitive matrix $C_{S^{p,n}, D^{p,n}}$ and $C_{D^{p,n}, D^{p,n}}$ are

$$C_{S^{p,n}, D^{p,n}} = \frac{1}{N_e - 1} \sum_{j=1}^{N_e} (s_j^{p,n} - \bar{s}^{p,n})(d_j^{p,n} - \bar{d}^{p,n})^T \tag{10.4}$$

where $\bar{s}^{p,n} = \frac{1}{N_e} \sum_{j=1}^{N_e} (s_j^{p,n})$; $\bar{d}^{p,n} = \frac{1}{N_e} \sum_{j=1}^{N_e} (d_j^{p,n})$.

$$C_{D^{p,n}, D^{p,n}} = \frac{1}{N_e - 1} \sum_{j=1}^{N_e} (d_j^{p,n} - \bar{d}^{p,n})(d_j^{p,n} - \bar{d}^{p,n})^T \tag{10.5}$$

From Eq. (10.2), based on EnKF [13], history matching only needs restarting calculation in different time step for each reservoir model, instead of starting from initial time step of the simulator. Thus, approximately, the expense (simulation times) only relates to the number of models, which suits solving large-scale history matching problem [9–12].

10.3 Reservoir Production Optimization

For the production optimization problem [12], one wishes to achieve the maximum NPV or field cumulative oil production for an expected reservoir life time [13, 19–21]. We define the vector u as the control variable to be optimized, which contains the control parameters such as production rates, injection rates, and well bottom hole pressure et al.

$$u = [u_1, u_2, u_3, \cdots, u_{N_u}]^T \tag{10.6}$$

where N_u represents the total number of control variables, which equals to control well numbers times total number of control time step. Reservoir development economic benefits represent as following [16, 17, 22–24]:

$$J(u, s) = \sum_{n=1}^{N_t} \frac{\left[r_o Q_o^n(u, s) - r_w Q_w^n(u, s) - r_{wi} Q_{wi}^n(u, s) \right] \Delta t^n}{(1 + b)^{t^n}} \tag{10.7}$$

where J is the objective function which represents NPV [22]. N_t is the total number of control time step; r_o is the oil revenue (yuan/m^3); r_w and r_{wi} are, respectively, the water production and water injection costs (yuan/m^3); Q_o^n and Q_w^n are, respectively, the average oil and water production rates of the nth simulation time step; Q_{wi}^n is the average water injection rate of the nth water injector (m^3/d); b is the annual discount rate (%); Δt^n is the nth simulation time step (d) and t^n is the cumulative time up to the nth simulation time step (year). Q_o^n and Q_w^n are decided by control variable u and reservoir model s. Thus, J is the function of u and s.

In order to match the uncertainty of optimized strategy and reservoir model, based on robust optimization, the expectation of index function J is considered as optimization objective function to obtain maximization. By EnKF, updated N_e number model realizations as foundation [16, 17, 22], problem can be described as following:

$$Max\ J_S(u) = \frac{1}{N_e} \sum_{j=1}^{N_e} J(u, s_j) \tag{10.8}$$

where J_S is the average NPV calculated by every model realization; $J(u, s_j)$ represents the NPV of the jth model realization.

Because of the extreme difficulty of obtaining gradient of J_S which relates to control vector u, SPSA is adopted. Gradients can be obtained by perturbating all control variables at the same time. Although perturbation gradients are stochastic, research uphill direction can be guaranteed for maximization problem [9]. Obviously, J_S and $J(u, s_j)$ have linear relationship, perturbation gradients of J_S can be represented by stochastic gradients of index function in a linear way. At the lth iteration, stochastic gradient $\hat{g}_S^l$ of J_S represents as

$$\hat{g}_S^l(u^l) = \frac{1}{N_e} \sum_{j=1}^{N_e} \hat{g}_j^l(u^l) \tag{10.9}$$

where u^l is the optimized control vector at the lth step; $\hat{g}_S^l$ is the stochastic perturbation gradient of $J(u, s_j)$ which represents as

$$\hat{g}_j^l(u^l) = \frac{J\left(u^l + \varepsilon_l \Delta_l^j, s_j\right) - J\left(u^l, s_j\right)}{\varepsilon_l} \times \Delta_l^j \tag{10.10}$$

where ε_l is the perturbation step size; Δ_l^j is a dimensional stochastic perturbation vector in which parameters from symmetric ± 1 Bernoulli or standard normal distribution are totally independent. Ultimately, optimized result shows great volatility which makes it difficult to analyze the regularity of optimization control. Therefore, control variable covariance matrix is taken into consideration and creates related Gauss stochastic vectors to calculate perturbation gradients which represent as

$$\hat{g}_j^l(u^l) = \frac{J\left(u^l + \varepsilon_l C^{1/2} z_l^j, s_j\right) - J\left(u^l, s_j\right)}{\varepsilon_l} \times C^{1/2} z_l^j \tag{10.11}$$

where z_l^j is a vector of independent normal deviates with zero mean and unit variance, i.e., $N(0, 1)$; C is the control variable covariance matrix; $C^{1/2}$ is the square root matrix of C, which can be obtained by Cholesky decomposition method and satisfies $C^{1/2}C^{T/2} = C$; $C^{1/2}z_l^j$ is a Gaussian stochastic perturbation vector which obeys standard normal distribution with zero mean and C variance. It is usually assumed that C as spherical or Gaussian model and here adopts spherical model [1, 12].

According to the Taylor's series expansion, the function $J\left(u^l + \varepsilon_l C^{1/2}z_l^j, s_j\right)$ becomes

$$J\left(u^l + \varepsilon_l C^{1/2}z_l^j, s_j\right) = J\left(u^l, s_j\right) + \varepsilon_l (C^{1/2}z_l^j)^T g_j^l(u^l)$$
$$+ o\left(\left\|\varepsilon_l C^{1/2}z_l^j\right\|^2\right) \tag{10.12}$$

where $g_j^l(u^l)$ is the real gradient of $J(u, s_j)$ at the point of u^l. Neglect the third term and substitute Eq. (10.12) into Eq. (10.11) yields

$$\hat{g}_j^l(u^l) = C^{1/2}z_l^j (C^{1/2}z_l^j)^T g_j^l(u^l)$$
$$= C^{1/2}z_l^j (z_l^j)^T C^{T/2} g_j^l(u^l) \tag{10.13}$$

Substitute $\hat{g}_j^l(u^l)$ into Eq. (10.9) and then obtain expectations of both sides,

$$E[\hat{g}_S^l(u^l)] = \frac{1}{N_e} \sum_{j=1}^{N_e} E[\hat{g}_j^l(u^l)]$$

$$= \frac{1}{N_e} \sum_{j=1}^{N_e} C^{1/2} E[z_l^j (z_l^j)^T] C^{T/2} g_j^l(u^l)$$

$$= \frac{1}{N_e} \sum_{j=1}^{N_e} C^{1/2} I_{N_u} C^{T/2} g_j^l(u^l) \tag{10.14}$$

$$= C \frac{1}{N_e} \sum_{j=1}^{N_e} g_j^l(u^l) = C g_S^l(u^l)$$

where I_{N_u} is unit matrix of N_u dimension; $\hat{g}_j^l(u^l)$ is the real gradient of $J(u, s_j)$ at the point of u^l.

The expectation of perturbation gradient calculated by Eq. (10.14) is the product of the control variable covariance matrix and the real gradient. Therefore, the calculated search direction can be approximately viewed as the quasi-Newton direction of inverse Hessian matrix replaced by the covariance matrix. Moreover, C is generally a positive definite matrix so that the convergence of the algorithm can be guaranteed.

After obtaining perturbation gradient, at the $(l + 1)$th iteration with line search method yields, control parameter is

$$u^{l+1} = u^l + \alpha_l \frac{\hat{g}_S^l(u^l)}{\left\| \hat{g}_S^l(u^l) \right\|_\infty} \tag{10.15}$$

where α_l is a trial step size; $\left\| \hat{g}_S^l(u^l) \right\|_\infty$ means the infinity norm of the vector. The step size α_l is determined by inexact line research, where α_l will be cut by half if objective function is not increased strictly at the present step. This step will not stop repeating until the objective function yields a strict increase. For convergence criteria, we use both of the following two conditions, i.e.,

$$\frac{\left| J_S \left(u^{l+1} \right) - J_S \left(u^l \right) \right|}{J_S \left(u^l \right)} \leq 3 \times 10^{-3} \tag{10.16}$$

And

$$\frac{\left\| u^{l+1} - u^l \right\|}{max(\left\| u^l \right\|, 1.0)} \leq 3 \times 10^{-3} \tag{10.17}$$

10.4 Computing Process

Reservoir numerical simulation has become an effective technology that can reproduce the entire development process of the oilfield and make future predictions. In order to verify the effectiveness of the method proposed above, combined with reservoir numerical simulation technology, the calculation process for reservoir closed-loop production optimization management is given as followed.

1. Generating a set of initial reservoir model realizations based on prior data, setting control parameter vectors of wells according to production situation, initial models begin to optimization by the SPSA method, making sure the optimized strategies under uncertainty.
2. Substitute obtained develop strategy into real reservoir model and run the simulation process, thus production observed data at the first step will be created. This process represents reality of reservoir production.
3. Based on the first step observed data, initial model realization updating and verification by the EnKF method, in order to decrease the uncertainty of reservoir model.
4. Based on updated model realization, the next step is optimization and data assimilation. Settle down the optimized strategy and further reduce uncertainty, keeping previous program until closed-loop complete.

10.5   Example Application and Analyzation

First step, establishing a 2D synthetic reservoir model with three phases which has a uniform grid system with $20 \times 30 \times 1$ grid blocks and $\Delta x = \Delta y = 45\,\text{m}$, $\Delta z = 9.1\,\text{m}$. Figure 10.2 shows the real permeability distribution of reservoir. The initial pressure is 17.2 Mpa and the initial oil saturation is 0.8 for the whole reservoir. There are a total of 13 wells which include 4 producers and 9 injectors arranged by 4 five-spot well patterns. The anticipated waterflooding project life is 1200 days and the control time step size is set to 120 days, so there are 10 control steps and $(4 + 9) \times 10 = 130$ control variables. During optimization [18], all the injectors are under injection rate control with an upper bound of $170\,\text{m}^3$/day and a lower bound of $0\,\text{m}^3$/day as well as the producers are under BHP control with the upper bound and lower bound of 25 and 12 MPa, respectively. The oil price is 2800 yuan/m^3, the water production cost is 220 yuan/m^3, and the annual discount rate is 5%. The observation data used for history matching are bottom hole pressure (BHP) of injectors, water production rate (WPR), and oil production rate (OPR).

A closed-loop reservoir management test was performed on this synthetic reservoir model based on the proposed method. Initial model realizations created by sequential Gaussian simulation, total number $N_e = 50$. During the closed-loop management, NPV at different iterations based on initial realizations is shown in Fig. 10.3. The black bold curve with marked points means average change of NPV. The black curve means change of NPV for each model realization. Obviously, NPV of every model realization gets a notable improvement after optimization. In the way of probability, production optimization strategy is more reliable and reduces the dependability to reservoir model, thus decreases the risk of development.

Figure 10.4 represents the updated porosity of average reservoir model by EnKF, which inverses the high porosity area precisely after the 720 days (sixth step) and average porosity of the model maintains well. Figures 10.5 and 10.6 present

Fig. 10.2 Field log horizontal permeability distribution

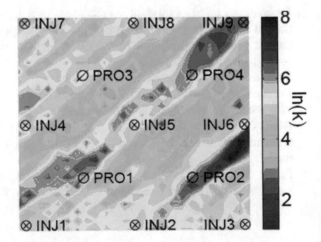

Fig. 10.3 The calculated
NPV at different iterations
based on initial realizations

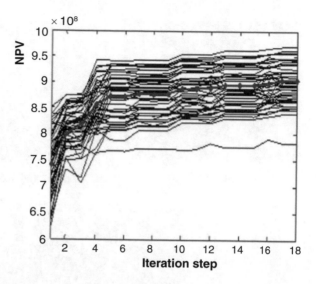

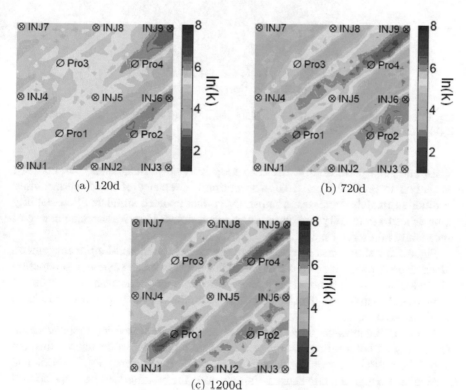

Fig. 10.4 The updated horizontal log permeability of mean model at various time. (**a**) 120 days.
(**b**) 720 days. (**c**) 1200 days

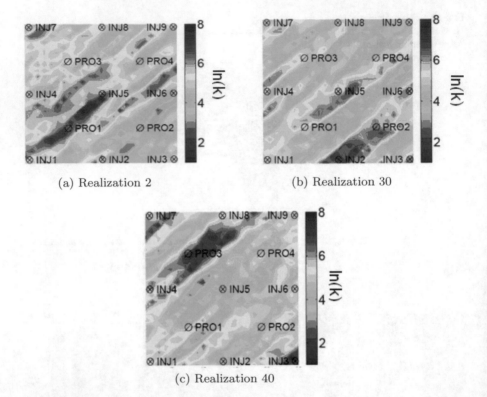

(a) Realization 2   (b) Realization 30

(c) Realization 40

Fig. 10.5 The initial horizontal log permeability distribution of some realizations. (**a**) Realization 2. (**b**) Realization 30. (**c**) Realization 40

initial and final permeability of closed-loop reservoir management, respectively. According to Figs. 10.5 and 10.6, distribution discrepancy of initial permeability in different models decreases prominently so that updated simulated permeability approaches to the reality and closed-loop management improves understanding of uncertainty in reservoir model.

Figure 10.7 shows part of data assimilation results in closed-loop management. Black discrete points mean observed data. Black curve means dynamic production data calculated by each model realization. Gray bold curve means average value of previous calculated data. Through constant assimilation, predict data could match the observed data.

Figure 10.8 shows the final optimal controls through closed-loop management. The change of color reflects the change of control variable parameters in different steps. Well PRO2 maintains high BHP in most of the time; INJ2, INJ3, and INJ8 maintain high injection speed. Furthermore, the relationships between control parameters are taken into consideration, the optimal controls become smoother which could manipulate easily in oilfield.

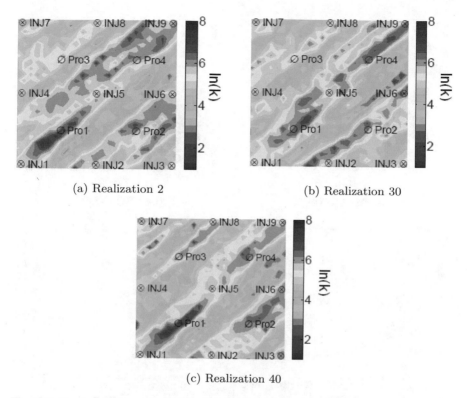

(a) Realization 2   (b) Realization 30

(c) Realization 40

Fig. 10.6 The final horizontal log permeability distribution of some realizations. (**a**) Realization 2. (**b**) Realization 30. (**c**) Realization 40

Substituting closed-loop production optimization management into reservoir model, computing residual oil distribution by simulation, and comparing to reactive control results that are shown in Fig. 10.9. Figure 10.10 shows the changes of field water cut of two kinds of strategies. According to Figs. 10.9 and 10.10, optimized water sweep efficiency increases remarkable, field water cut lower than reactive control during well production life and NPV achieves.

10.6   Conclusion

1. The new method of multiple model production optimization based on SPSA decreases the uncertain sensitivity of optimization to reservoir model.
2. Considering the introduction of the control variable covariance matrix, the optimal controls calculated by the SPSA algorithm is more smooth and continuous. The calculated search direction is similar to the quasi-Newton direction of inverse Hessian matrix using the covariance matrix.

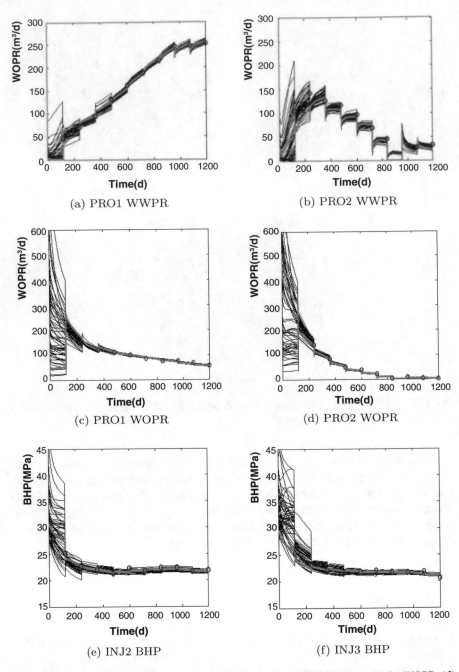

Fig. 10.7 The data matching result. (**a**) PRO1 WWPR. (**b**) PRO2 WWPR. (**c**) PRO1 WOPR. (**d**) PRO2 WOPR. (**e**) INJ2 BHP. (**f**) INJ3 BHP

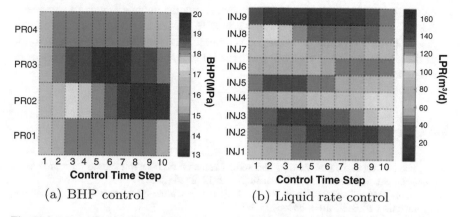

(a) BHP control

(b) Liquid rate control

Fig. 10.8 The optimal well controls obtained by closed-loop management. (**a**) BHP control. (**b**) Liquid rate control

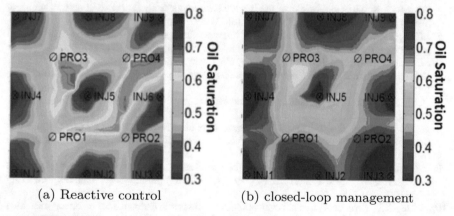

(a) Reactive control

(b) closed-loop management

Fig. 10.9 Comparison of remaining oil saturation distribution. (**a**) Reactive control. (**b**) Closed-loop management

Fig. 10.10 Comparison of field water cut

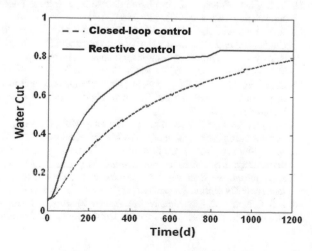

3. According to the result, closed-loop reservoir management enhances water sweep coefficient remarkable and NPV increase 23% compared to reactive control. And geological model estimation can be obtained iteratively without the calculation of the adjoint gradient, which is quite useful to be coupled with any commercial simulator.

References

1. H. Zhao, Y. Li, J. Yao, K. Zhang, Theoretical research on reservoir closed-loop production management. Sci. China Technol. Sci. **54**(10), 2815–2824 (2011)
2. Y. Chen, D.S. Oliver, D. Zhang et al., Efficient ensemble-based closed-loop production optimization. SPE J. **14**(04), 634–645 (2009)
3. C. Chen, Y. Wang, G. Li, A.C. Reynolds, Closed-loop reservoir management on the Brugge test case. Comput. Geosci. **14**(4), 691–703 (2010)
4. L. Peters, R. Arts, G. Brouwer, C. Geel, S. Cullick, R.J. Lorentzen, Y. Chen, N. Dunlop, F.C. Vossepoel, R. Xu et al., Results of the Brugge benchmark study for flooding optimization and history matching. SPE Reserv. Eval. Eng. **13**(03), 391–405 (2010)
5. J.C. Spall et al., Multivariate stochastic approximation using a simultaneous perturbation gradient approximation. IEEE Trans. Autom. Control **37**(3), 332–341 (1992)
6. Æ. Geir, T. Mannseth, E.H. Vefring et al., Near-well reservoir monitoring through ensemble Kalman filter, in *SPE/DOE Improved Oil Recovery Symposium* (Society of Petroleum Engineers, 2002)
7. H. Zhao, Y. Li, Z. Kang, Robust optimization in oil reservoir production. Acta Petrol. Sin **34**(5), 947–953 (2013)
8. H. Zhao, C. Chen, S.T. Do, G. Li, A.C. Reynolds et al., Maximization of a dynamic quadratic interpolation model for production optimization, in *SPE Reservoir Simulation Symposium* (Society of Petroleum Engineers, 2011)
9. Y. Gu, D.S. Oliver, The ensemble Kalman filter for continuous updating of reservoir simulation models. J. Energy Res. Technol. **128**(1), 79–87, 07 (2005) [Online]. Available https://doi.org/10.1115/1.2134735
10. N. Liu, D.S. Oliver et al., Critical evaluation of the ensemble Kalman filter on history matching of geologic facies, in *SPE Reservoir Simulation Symposium* (Society of Petroleum Engineers, 2005)
11. Y. Gu, D.S. Oliver et al., An iterative ensemble Kalman filter for multiphase fluid flow data assimilation. SPE J. **12**(04), 438–446 (2007)
12. A.A. Emerick, A.C. Reynolds et al., EnKF-MCMC, in *SPE EUROPEC/EAGE Annual Conference and Exhibition* (Society of Petroleum Engineers, 2010)
13. E. Aliyev, L.J. Durlofsky, Multilevel field development optimization under uncertainty using a sequence of upscaled models. Math. Geosci. **49**(3), 307–339 (2017)
14. C. Wang, G. Li, A.C. Reynolds et al., Production optimization in closed-loop reservoir management, in *SPE Annual Technical Conference and Exhibition* (Society of Petroleum Engineers, 2007)
15. G. van Essen, M. Zandvliet, P. Van den Hof, O. Bosgra, J.-D. Jansen et al., Robust waterflooding optimization of multiple geological scenarios. SPE J. **14**(01), 202–210 (2009)
16. M.G. Shirangi, L.J. Durlofsky et al., Closed-loop field development optimization under uncertainty, in *SPE Reservoir Simulation Symposium* (Society of Petroleum Engineers, 2015)
17. J.-D. Jansen, R. Brouwer, S.G. Douma et al., Closed loop reservoir management, in *SPE Reservoir Simulation Symposium* (Society of Petroleum Engineers, 2009)
18. D.S. Oliver, A.C. Reynolds, N. Liu, *Inverse Theory for Petroleum Reservoir Characterization and History Matching* (Cambridge University Press, Cambridge, 2008)

19. Y. Li, X. Liu, X. Chen, Optimization of large-scale water flooding system. Acta Petrol. Sin. **22**(6), 69–72 (2001)
20. L.J. Durlofsky, R. Dimitrakopoulos, Smart oil fields and mining complexes. Math. Geosci. **49**(3), 275–276 (2017)
21. W. Sun, L.J. Durlofsky, A new data-space inversion procedure for efficient uncertainty quantification in subsurface flow problems. Math. Geosci. **49**(6), 679–715 (2017)
22. H.X. Vo, L.J. Durlofsky, Data assimilation and uncertainty assessment for complex geological models using a new PCA-based parameterization. Comput. Geosci. **19**(4), 747–767 (2015)
23. O. Volkov, D. Voskov, Effect of time stepping strategy on adjoint-based production optimization. Comput. Geosci. **20**(3), 707–722 (2016)
24. M.G. Shirangi, O. Volkov, L.J. Durlofsky et al., Joint optimization of economic project life and well controls. SPE J. **23**(02), 482–497 (2018)

Chapter 11
The Optimization of Maritime Search and Rescue Simulation System Based on CPS

Lin Mu and Enjin Zhao

11.1 Introduction

In recent years, a series of major maritime accidents have caused serious casualties and economic losses. It is urgent to deal with maritime emergencies properly. After the accident happened in the sea, the most urgent task is to quickly and accurately locate the rescue target or determine the rescue area where the target is in distress.

The drift of the distress target is a dynamic time-varying process. To locate the position of the target in distress accurately, the ideal method is to establish communication with it. However, due to the limitation of the actual situation and technical skill, the current maritime search and rescue methods still adopt conventional methods such as the blanket search or the search depending on the previous search experience, which spends a lot of manpower and material resources for blindly searching the whole ocean of wreck. Therefore, prime time for search and rescue might be wasted since it is difficult to determine the search area efficiently and effectively using the existing maritime search and rescue decision-making system.

Due to the emergence of the maritime search and rescue requirement and the urgency of the processing, commands and decisions must be made thoroughly as soon as possible. However, related real-time data need to be sufficient, such that scientific commands and decisions can be determined. Therefore, developing an effective way of gathering, sharing, and comprehensively processing data from various sources has become the key to improving the performance of the maritime

L. Mu (✉)
College of Marine Science and Technology, China University of Geosciences, Wuhan, People's Republic of China

E. Zhao
Shenzhen Research Institute, China University of Geosciences, Shenzhen, People's Republic of China

© Springer Nature Switzerland AG 2020
S. Hu, B. Yu (eds.), *Big Data Analytics for Cyber-Physical Systems*,
https://doi.org/10.1007/978-3-030-43494-6_11

search and rescue nowadays so that decision-making can be intuitively and visually assisted. It is also the main difference between the modernized emergent commanding based on informatization and intelligentization and the traditional emergent commanding based on regional, departmental, and empirical closure methods.

With the fast development of computational sciences, a new proposed prediction model (MSRSS) [1], which has shown its advantages of rapidity and efficiency, is widely put into use in departments of maritime search and rescue. The model is able to calculate the drifting trajectory and predict the further trace of the target in distress according to the existing measurements by using a data-inversion way. The main characteristics of the model lie in three aspects: (1) Intuitiveness. The information of the distress, the search and rescue force, and the scene can be precisely positioned spatially; (2) Rapidity and intelligence. Depending on the modern computer technology, communication technology, and information processing technology, once the relevant parameters and the meteorological and hydrological conditions are defined, information including graphs and the data of target in distress can be automatically generated fast to assist decision-making; (3) Capability of simulating and replaying. All the decision-making procedures of search and rescue events can be recorded, processed, replayed, and analyzed repeatedly using numerical simulation, in order to conclude the optimal model of handling distress with similarities.

The core algorithm SSRF (Solver of Search and Rescue Formula) of the prediction model is the solution of the embedded formulas based on the boundary conditions which in practice are stored in the database. Simulation results can be predicted based on the core algorithm; however, the embedded formulas contain a large number of empirical parameters which need to be adjusted and validated properly to guarantee the accuracy and optimize of the model. The model optimization process has attributes of the cyber physical system (CPS) [2–5]. As shown in Fig. 11.1, various types of physical quantities are collected and transformed into analog quantities through corresponding sensors firstly. In this stage, all data collected by sensors are saved in the system. After that, the saved data are extracted as the basis of meteorological and hydrological conditions for numerical simulations. Based on it, the scope of search and rescue are able to be predicted by computational facilities. In the end, the accuracy of the model can be verified and optimized by comparing the difference between numerical results and experimental measurements. Iteratively looping the above procedure, the accuracy of the model can be improved and meet the requirement of the prediction eventually. The optimized model based on the cyber physical system is of vital importance to the decision-making of maritime search and rescue and the positioning of the target in distress [6].

Fig. 11.1 Sketch of the model optimization based on the cyber physical system

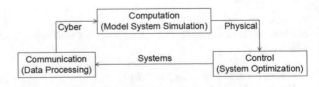

The marine environment prediction model plays an important role in the decision-making system of marine search and rescue, which includes the prediction drift trajectory of the search target, the survival time of the surviving person, and the detection ability of the search unit. However, the prediction accuracy and resolution of the marine environment prediction model are generally imperfect at present. Using CPS technology can further improve the prediction accuracy and horizontal resolution of the existing ocean forecasting models. The results of the model prediction can provide an important reference for the search and rescue decision. After the optimization of the MSRSS based on the CPS, the prediction accuracy of the model is improved, which is contributing to the marine safety guarantee.

This article explains the optimization procedure of the maritime search and rescue model in detail, organized as follows. The second section introduces the model of maritime search and rescue. The third section gives the procedure of the model optimization. The fourth section provides an example to illustrate the optimization process of the model. The fifth section lists conclusions.

11.2 Maritime Search and Rescue Model in MSRSS

In MSRSS, the hydrodynamic environments of the maritime search and rescue model are simulated depending on the weather research and forecasting model (WRF) and the finite volume coastal ocean model (FVCOM) [7], while the targets in distress are modeled using the Lagrangian particle tracking method. The LEEWAY model is used to determine the wind-induced divergence angle

The vertical coordinate of the WRF model is defined as the terrain-following static pressure coordinate. Main modules of the WRF model are listed as follows:

(1) WRF Preprocessing System (WPS) module, which is used to provide the background field of the simulation, initializes the area of the numerical simulation and sets the difference of the topographic data and the meteorological data (data from other patterns such as the Global Pattern) into the simulation area.
(2) WRF Data Assimilation (WRFDA) module improves the initial and boundary conditions needed in the simulation by applying scheme in assimilating objective data from observation stations, satellites, and radars.
(3) Main Simulation Program module generates the initial background field and the time-varying boundary conditions needed in the simulations to calculate formulas through numerical integration.
(4) Post-Processing module analyzes the output data of the model and illustrates the result graphs.

It is well known that precise predicted results rely highly on the quality of the initial field. In this model, single-step 3DVAR is used to assimilate data into the initial field of the time-varying prediction, however, on condition that the convergence rate of grids is high and the observation data are big enough, multi-

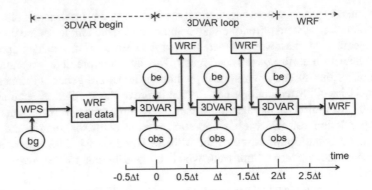

Fig. 11.2 Loop 3DVAR process

step 3DVAR (looping 3DVAR process) can be applied to form an initial field with higher quality, shown in Fig. 11.2.

In the figure, the first initial field is generated by the WRF preprocessing system (WPS) and real data, while WRF data drives the forward-time integration of WRF. "bg," "obs," and "be" represent the background field, observation field, and covariance between the observation field and the background field. "0," "Dt," and "2Dt" represent the assimilation moment, for example, assuming the assimilation window at a time Dt to be 0.5 Dt–1.5 Dt, all the observation data from this period are assimilated at the time Dt. FVCOM is used as the hydrodynamic prediction method which adopts integration of the volume flux to solve the governing equations of the fluid domain. Mass conservation is satisfied in every single grid of the whole domain, which is one of the important laws in ocean-related numerical simulations. Wet-dry boundary judgment method is applied in FVCOM to deal with boundary moving of the tidal flats.

The greatest feature and advantage of FVCOM in dealing with offshore and coastal events lie in the combination of finite element method and finite difference method. The finite element method adopts triangular grids and linearly independent basis functions, which is predominant in the boundary fitting and local refining, while the finite difference method discretizes the original fluid governing equations directly, which is predominant in the certainty of hydrodynamic theory, intuitiveness of difference, and efficiency of calculation. FVCOM has merits of both methods, which applies the integral forms of governing equations and better formats of computation to guarantee the better conservation of mass, momentum, and energy.

Governing equations of FVCOM contain the mass conservation equation, momentum conservation equation, temperature equation, salinity equation, and density equation. In the MSRSS, only the hydrodynamic modules are considered, which include the mass conservation equation and momentum conservation equation. The mass conservation equation which is also called the continuity equation is written as

$$\nabla \mathbf{u} = 0 \qquad (11.1)$$

The momentum conservation equation is written as

$$\frac{\partial \mathbf{u}}{\partial t} + (\mathbf{u} \cdot \nabla)\mathbf{u} = \mathbf{f} - \frac{1}{\rho}\nabla\rho + \nu\nabla^2\mathbf{u} \tag{11.2}$$

where ∇, u, ρ, f, ρ are the Hamiltonian operator, fluid velocity vector, the density of sea-water, mass force, and fluid pressure, respectively.

Comprehensively considering the force of sea surface wind on the drifting target, LEEWAY model proposes the concept of wind-induced divergence angle. The wind-induced drift is decomposed into two components, which are parallel to the wind direction and perpendicular to the wind direction. In the search and rescue system, nine parameters fully describe the characteristics of the drifting path, including the slope in the wind direction, offset in the wind direction, statistical fluctuation in the wind direction, slope in the right direction of the wind, offset in the right direction of the wind, statistical fluctuations in the right direction of the wind, slope in the left direction of the wind, offset in the left direction of the wind, and statistical fluctuations in the left direction of the wind. Additionally, the motion of the target corresponding to the marine meteorological environment can be more accurately portrayed by considering the force of surface currents on it. The velocity decomposition diagram of LEEWAY drift vector is shown in Fig. 11.3.

In Fig. 11.3, L_{drift} is the LEEWAY drift vector, while the two decomposed components of L_{drift} are L_d and L_c. The relationship between the LEEWAY drift vector and wind velocity is shown in Eq. (11.3):

$$L_{c,d} = a_{c,d}W_10 + b_{c,d} \tag{11.3}$$

where $a_{c,d}$ is the wind drift coefficient or LEEWAY factor determined from the experimental data.

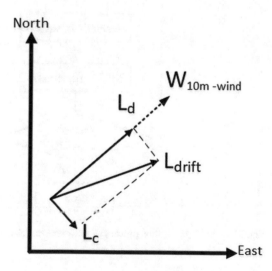

Fig. 11.3 Velocity decomposition diagram of wind

Lagrangian particle tracking method is adopted to predict the behavior and drift of the target in distress based on the hydrodynamic environmental parameters provided by the WRF and FVCOM. The prediction of the drifting traces on the ocean surface employs the Monte-Carlo random statistic theory, which considers the uncertainty of time and position of persons and goods in distress to define the initial search and rescue area, and generates several randomly distributed floaters within the initial area. For any single drifting floater, drifting positions and search and rescue areas with random statistical characteristics are obtained through the prediction of drifting traces by considering the uncertainty of drifting directions affected by wind (deviations from the wind direction) and the error of forecasted wind speed.

11.3 The MSRSS Model Optimization Process

The optimization process of maritime search and rescue simulation system based on CPS is shown in Fig. 11.4.

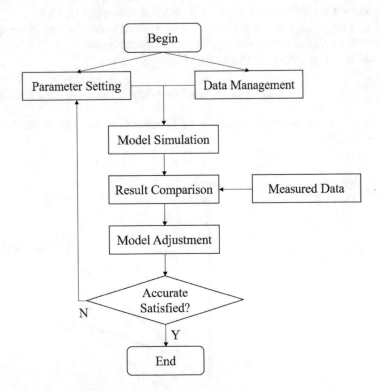

Fig. 11.4 The optimization process of maritime search and rescue simulation system

At the connection end of the system, the marine, meteorological, and hydrological data, forecast data from the global spectral model, NEAR_GOOS-SST, conventional ground and sounding observation data, and other data requested by the model operation are input, decoded, quality-controlled, and formatted. The multi-source observation data are performed by using variational assimilation technology. The system drives the WRF and FVCOM models using the above data to complete the calculation of equations, through which the simulation results are obtained and saved. In the meanwhile, the wind field, flow field, and the time-varying search and rescue range of the drifting trajectory based on Monte-Carlo's particle tracking method is physically measured during the simulation time. The accuracy of the prediction system is verified by comparing results between simulations and measurements. According to the simulation results, parameters in the system are adjusted and controlled to optimize the system. Assimilation of the optimized system and measured data, the data-collecting process, and simulation of the prediction model, etc., are repeated to complete a new cycle of system optimization until the accuracy of the simulation meets the calculation requirements [8]. Detailed steps of the optimization of maritime search and rescue system are as follows:

11.3.1   Data Management

1. Data classification:

   (a) Marine, meteorological, and hydrological data: temperature, salinity, flow field about velocity and direction, tide, wind field, temperature, pressure, etc.;
   (b) Marine basic geographic data: digital raster map, marine digital orthophoto map, digital elevation data, and terrain data;
   (c) Marine disaster reduction data: the impact assessment data of sea level change, benchmark tidal level verification and leveling data, marine tide forecast product data, and forecast inspection product data;

2. Data collection methods:

   (a) Near-shore currents, waves, sea ice, and meteorological data are collected by observation of shore-based radar;
   (b) Tide levels are collected by tide gauges of tide observation stations at port and dock;
   (c) The marine and meteorological elements in ocean–atmosphere interaction are collected by ocean weather station;
   (d) Ocean observations are undertaken by the scientific expedition ships, floats, satellites, volunteer ships, etc. The observations are focused on marine science, climate change, and ocean–atmosphere interaction.

3. Data repository:

   (a) Search and rescue resource database;
   (b) Basic and special geographic information database;
   (c) Dynamic ship and aircraft resource database connected with China Ship Reporting Center (CHISREP), Vessel Traffic Services Center (VTS), and Air Traffic Control Center (ATC);
   (d) Positional information database of marine perils;
   (e) Database of various types of search and rescue plans;
   (f) Weather and sea state information database;
   (g) Dynamic mathematic models to predict possible areas for drift and pollution diffusion;
   (h) Satellite remote sensing and aerial remote sensing information (Monitor oil spill at sea).

4. Data display:

   Based on three-dimensional sphere and integrated numerical analysis, the visualization of multi-dimensional profile of marine elements, such as ocean temperature, salinity, density, current, the display of three-dimensional scenes on the seabed, water, and island coast, the simulation of storm surge, sea level rise, and coast change and other phenomena can be realized.

5. Data processing and transmission:

   In order to ensure the reading of the data required by the next loop of model simulation, the collected data needs to be processed and stored in a readable format for the MSRSS model.

11.3.2   Model Simulation and Prediction

In the prediction model, the calculation methods include Lagrangian particle tracking method, the convex hull algorithm in Monte-Carlo theory, and topological geometry. The process of the model simulation is shown in Fig. 11.5.

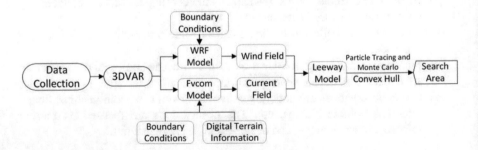

Fig. 11.5 The process diagram of model simulation

1. WRF model.

WRF model is a fully compressible and non-hydrostatic model, the control equations of which are written as flux forms. Predicted data from the global spectral model are read, extracted, objectively analyzed, and interpolated to create the initial field and boundary conditions required by the WRF model.

The WRF model mainly consists of the WRF preprocessing system (WPS), WRF data assimilation (WRFDA), main simulation program, and post-processing modules, which relies on three-dimensional variational data assimilation (3DVar) to assimilate meteorological observation data for obtaining a more accurate initial condition. According to the search and rescue requirement, WRF model provides outputs such as forecast wind field as the initial forced conditions for the FVCOM model.

2. FVCOM model.

FVCOM model reads the WRF prediction results directly using the Fortran program and the wind field output from WRF as a forcing condition. Depending on the simulation of the FVCOM model, different resolution data for the target in distress are provided. General steps of data extraction from WRF to FVCOM are listed as follows:

(a) Extracting sea surface wind data from the WRF forecast results.
(b) Translating sea surface wind data from Lambert projection to regular latitude and longitude coordinates.
(c) Processing the sea surface wind data in the latitude and longitude grids with NetCDF format.

The interconnection between modules controlled by Linux/UNIX shell programs is fully automatic. Combined with initial boundary conditions such as tides and terrain, FVCOM model provides the necessary flow field information for the trajectory prediction of the target in distress.

3. Lagrangian particle tracking and search and rescue range predicting

Based on the environmental dynamic fields from WRF and FVCOM associated with LEEWAY model, particle model is used to predict the drifting process of the target in distress. The prediction model of drifting trajectory in the sea surface uses the Monte-Carlo random statistic theory to define the initial search range, with uncertainties such as the initial time and location of persons or goods falling into the water taking into account. The convex hull algorithm is used to determine the search range over time according to the trend of the initial distribution of particles.

It should be mentioned that the search and rescue system provides 13 h's prediction results of drifting trajectory and search range.

11.3.3 Comparison Between Measured Data and Simulation Data

The simulation results of the model are compared with the measured data, while the reasons for the difference between the two results are analyzed. Depending on the analysis, model parameters or empirical formulas are adjusted to optimize the model. The comparison between the predicted results and the measured data is as follows:

(1) Tidal harmonic constant.

The prediction harmonic constants are compared with the observation harmonic constants, with the amplitude error and mean late angle error between observational data and simulated data analyzed. These errors may be mainly caused by two aspects. On the one hand, the horizontal grid spacing of the model ranges from tens meters to thousands of meters, thus the resolution of the grids is not high in the vicinity of the coastline. The locations of grid-points in the model are different from the locations of tide gauge stations, leading to errors. On the other hand, the change of the shore boundary and the error of topography and water depth are responsible for the errors. In addition, it is possible that the errors might be related to the harmonic constant precision and other conditions as well.

(2) Tidal current.

The tidal current values from model simulations and measurements are compared, including tidal flow velocity and tidal flow direction. Meanwhile, causes of the error are analyzed, which lay the foundation for model adjustment.

(3) Water elevation.

The water elevations at verification points are obtained and compared with the measured data. The relative error, the root mean square (RMS), and the correlation coefficient between the simulated values and the measured values are analyzed. It is specified that usually under condition of the RMS error to be basically within 0.1 cm; the ratio of the observed deviation and the mean value to be less than 10%; and the correlation coefficient to be more than 0.9, could the simulation results of the water level be deemed as accurate, otherwise, the model should be adjusted.

(4) Track and scope of search and rescue model.

Numerical simulation of the prediction model is carried out on the trajectory and scope of the target in distress, while the accuracy of the simulation is analyzed by comparing the prediction results with the test results. If the distance between the simulation location and the experimental location is within 3 km, the prediction accuracy of the model can meet the requirements.

11.3.4   Model Adjustment

There are significant random features in the motion of objects drifting on the sea. A large number of empirical parameters are introduced into the model calculation. Moreover, several probability formulas are also introduced to the model operation to deal with the uncertain factors. Therefore, in order to improve the precision of the simulation, it is necessary to find the governing equations in the model based on the comparison of the results. The parameters and variables associated with the governing equations are analyzed, while various empirical parameters are optimized. By adjusting different parameters that impact the trajectory of objects, the probability density function which is part of the Monte-Carlo theory of the position of the object can be improved.

According to the comparison of tidal harmonic constant, tidal current, water level, search and rescue model track and range, the influence parameters are mainly included in the several aspects:

In the force equation of a drifting object, the time step parameter of the model depends on the time and space of the force (wind and flow) acting on the object. The control parameters of the influence of surface laminar flow on the drifting path: the real sea surface laminar flow is composed of various frequencies, and the combined control parameters of different frequency flows have an important influence on the prediction of the simulated particle trajectories.

The wind drift coefficient $a_{c,d}$ denotes the wind force on the drifting path. The empirical relationship between wind drift and wind velocity appeared here is obtained through the fitting formulas of many years observation data, which needs to be adjusted and checked continuously.

The drifting path is under the combined action of wind and current. The second-order Runge–Kutta iterative algorithm is used in the combination of wind and flow, and the proportion of wind and sea current in this algorithm is controlled by different parameters.

The perturbation term of wind drift coefficient, wind velocity, and flow velocity in Lagrangian particle tracking trajectory calculation equation also need to be optimized constantly. Besides, particle swarm optimization (PSO) is used in the particle drifting optimization [9], which is a computational method that optimizes a problem by iteratively improving candidate solutions with regarding a given measure of solution quality. It solves a problem by having a population of candidate solutions, here dubbed particles, and moving these particles around in the search space according to the simple mathematical formula over the position and velocity of particles. Each movement of a particle is influenced by its local best-known position and is also guided toward the best-known positions in the search space, which are updated as better positions found by other particles. Note that in each loop of optimization, the model needs to be verified again until it gets the best solution [10].

11.4   The Example of Model Optimization

This section analyzes the whole optimization process of the MSRSS, taking the loss of Malaysia Airlines flight MH370 as an example [11]. On 8 March 2014, the flight MH370 disappeared when it flew from Malaysia to Beijing. On 29 July 2015, a piece of marine debris was found on Reunion Island in the South Indian Ocean.

Firstly, the observation data about the environmental condition are collected and managed. Based on the LEEWAY sea drift theory and the Monte-Carlo theory, the prediction model in the South India Ocean is established to predict the drifting trajectory of the target. Different drifting objects have different wind pressure characteristics, and the drift parameters of 63 common drifters are given in the LEEWAY drift model. However, the drift parameter of the aircraft flap is not given. In the empirical experiment, three different drift parameters of the flap are taken, which are 1.2%, 1.5%, and 1.8%, respectively. After the calculation, the simulation results and the observation data are compared. Depending on the solution analysis, the optimal drift parameter is selected to be used in the MSRSS.

In the optimization of the MSRSS, the mesh of the suspected crash area should be established with the resolution of 5°. The physical data of coastal lines, bathymetry, and boundary conditions are collected from NOAA and physical sensors. Various types of physical quantities are transformed into the analog quantities, which can be read by models of FVCOM and WRF. All the collected data are saved in the system. The coefficients of both models are set as the constants without the drift coefficients in the LEEWAY model. In the first simulation, the drift coefficient is set at 1.2%. 1024 drift targets were distributed in the crash zone. After that, the research model is working, and the drift trajectory of the distressed target is predicted. Comparing the zone of the simulation targets with that of real debris, the accuracy of the prediction model is verified. If the simulation scope of research and rescue is agreed well with the real area of the debris, the optimization of the simulation model is finished. Otherwise, the drift coefficient is set to the 1.5% and the model is adjusted. Depending on the saved collected data, the adjusted model is working again and predicts the rescue area. The second loop of the model optimization is implemented until the prediction accuracy of the model meets the requirements of the forecast.

Due to the fact that initial time of target release has a certain effect on the drift of the target, the drift targets are released at three different initial times, which are March 6, 2014, March 10, 2014, and March 14, 2014. The end time of the prediction is July 30, 2015. The time step of calculation is 1 h. The optimization simulation results of three experiments are shown in Fig. 11.6.

The final drift position of 9000 targets in each simulation is shown in Table 11.1. In the three simulations, the number of targets passing through the ocean near the Reunion Island is 540, 490, and 480, and the corresponding probability is 6 per hundred, 5.4 per hundred, and 5.3 per hundred, respectively. It shows that the calculation results are very close at different initial times. The number of targets with different wind drift parameters is statistically analyzed, revealing that when the wind

Fig. 11.6 The process diagrams of model simulation at three different times; (**a**) March 6, 2014; (**b**) March 10, 2014; (**c**) March 14, 2014

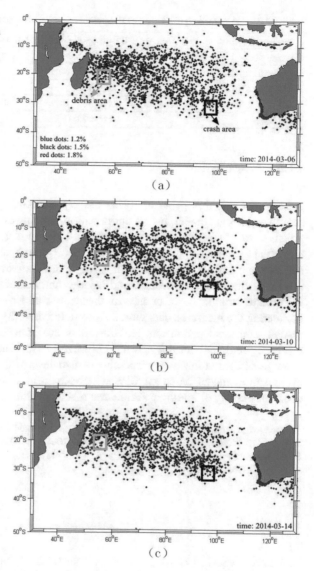

(a)

(b)

(c)

drift parameter is 1.8%, the accuracy of the prediction is the highest. Therefore, the drift parameter of 1.8% is selected and used in this MSRSS.

In short, the optimal wind drift coefficient is extracted with the CPS technique. Based on the wind drift coefficient, the maritime search and rescue system is optimized. The drift trajectory of the MH370 debris is predicted by the optimized system.

Table 11.1 The number (possibility) of the objects that finally reach Reunion Island area

Experiment condition	Exp. 1(2014-03-06)	Exp. 2(2014-03-10)	Exp. 3(2014-03-14)	Probability
Leeway factor 1.2%	110	50	80	2.6%
Leeway factor 1.5%	140	130	140	4.5%
Leeway factor 1.8%	290	310	260	9.5%
Total	540	490	480	
Probability	6%	5.4%	5.3%	

11.5 Conclusion

Based on CPS, this paper analyzes the optimization processing of the marine search and rescue model. In this optimization, sensing and communication systems are used to obtain the database required in the model. Helped by the power of the efficient computing server, the maritime search and rescue model quickly solves the inlay formula and predicts both the drift trajectory of the target and the range of search and rescue, and a large number of the calculation prediction results are stored. The accuracy of the search and rescue model is verified by comparing the measured data with the predicted data. By analyzing the verification results, empirical parameters and formulas are adjusted to optimize the model until the prediction accuracy of the optimized model meets the requirements. An example of the optimization processing is also described in the paper. The proposed optimization model based on CPS will be helpful to the implementation of the decision-making of maritime search and rescue in China.

Acknowledgments This work was supported by the National Key Research and Development Program of China (Grant No. 2017YFC1404700), the Discipline Layout Project for Basic Research of Shenzhen Science and Technology Innovation Committee (Grant No. JCYJ20170810103011913).

References

1. J. Gao, L. Mu, C. Li, G. Wang, J. Dong, J. Song, L. Han, Drifting trajectory analysis of debris from MH370 in the southern Indian Ocean, in *Oceans* (2016), pp. 1–4
2. H. Fawzi, P. Tabuada, S. Diggavi, Secure estimation and control for cyber-physical systems under adversarial attacks. IEEE Trans. Autom. Control **59**(6), 1454–1467 (2014)
3. E.A. Lee, Cyber physical systems: design challenges, in *IEEE International Symposium on Object Oriented Real-Time Distributed Computing* (2008), pp. 363–369
4. F. Pasqualetti, F. Dörfler, F. Bullo, Attack detection and identification in cyber-physical systems. IEEE Trans. Autom. Control **58**(11), 2715–2729 (2013)
5. L. Sha, S. Gopalakrishnan, X. Liu, Q. Wang, Cyber-physical systems: a new frontier, in *IEEE International Conference on Sensor Networks, Ubiquitous, and Trustworthy Computing* (2008), pp. 1–9

6. X. Chen, D. Zhang, L. Wang, N. Jia, Z. Kang, Y. Zhang, S. Hu, Design automation for interwell connectivity estimation in petroleum cyber-physical systems. IEEE Trans. Comput. Aided Des. Integr. Circuits Syst. **36**(2), 255–264 (2017)
7. C. Chen, J. Qi, C. Li, R.C. Beardsley, H. Lin, R. Walker, K. Gates, Complexity of the flooding/drying process in an estuarine tidal–creek salt–marsh system: an application of FVCOM. J. Geophys. Res. Oceans **113**(C07052), 1–21 (2008)
8. X. Chen, D. Zhang, Y. Wang, L. Wang, A. Zomaya, S. Hu, Offshore oil spill monitoring and detection: improving risk management for offshore petroleum cyber-physical systems: (invited paper), in *IEEE/ACM International Conference on Computer-Aided Design* (2017), pp. 841–846
9. J. Kennedy, R. Eberhart, Particle swarm optimization, in *International Conference on Neural Networks (ICNN'95)*, vol. 4 (2002), pp. 1942–1948
10. J. Wurm, Y. Jin, Y. Liu, S. Hu, *Introduction to Cyber-Physical System Security: A Cross-Layer Perspective*, vol. 3(3) (2017), pp. 215–227
11. J. Gao, M.U. Lin, G.S. Wang, L.I. Cheng, J.X. Dong, X.W. Bao, L.I. Huan, J. Song, Drift analysis and prediction of debris from Malaysia Airlines flight MH370. Chin. Sci. Bull. **61**(21), 2409–2418 (2016)

Chapter 12
A CPS-Improved Data Estimation Model for Flash Flood Early Warning Sensor Network

Zhanya Xu, Xiangang Luo, Shuang Zhu, Di Wu, and Qi Guo

12.1 Introduction

Flash floods are natural disasters in small areas that usually occur in basins of several hundred square kilometers or less. They typically occur within a few hours after heavy rain or short-term heavy precipitation [1]. As a result of steep slopes, impervious ground, saturated and moist soil, and some human factors, the water level in the basin changes rapidly with heavy rain. Flash floods have attracted increasing attention in the study of natural disasters [2, 3] owing to the large losses and the number of deaths associated with these natural disasters.

Because heavy rains often occur within hours of heavy rain, and the geographical environment of mountainous regions is mostly complex, disaster prediction is very difficult [4], and most disasters cannot be prevented [5]. Therefore, real-time online dynamic monitoring of surface runoff using sensor networks is an important tool for flash flood monitoring [6, 7]. In many countries, the implementation of early warning flash flood systems by deploying hydrological observatories in combination with hydrological models to determine hydrological thresholds has proven to be an effective means of reducing the losses due to mountain torrents [8]. Lightning, rainfall and landslides will affect the data quality and stability of sensors, thereby reducing the data service quality of the sensor network. The stable data service of the sensor network will provide reliable support for early warning of flash floods.

Z. Xu (✉) · X. Luo · S. Zhu · D. Wu
School of Geography and Information Engineering, China University of Geosciences, Wuhan, People's Republic of China

National Engineering Research Center for Geographic Information System, Wuhan, China

Q. Guo
Wuhan Tianhong Lightning Protection Testing Center Development Co., Ltd., Wuhan, China

© Springer Nature Switzerland AG 2020
S. Hu, B. Yu (eds.), *Big Data Analytics for Cyber-Physical Systems*,
https://doi.org/10.1007/978-3-030-43494-6_12

Cyber physical systems (CPS) has been a rapidly growing field in recent years [9]. It has great practicality in numerous areas where real-time, robust monitoring is urgently needed. With integrated management of various sensors and network resources, CPS is able to exchange information in a timely and reliable manner in different fields, providing reliable service by reconstructing computations and network abstractions [10]. Considering the advantages of CPS in real-time processing and robust protection, a data service framework for a flash flood warning sensor network is designed to provide fast and stable data service decision by detecting data anomalies in real-time.

Furthermore, in order to change the passive service mode of the existing sensor network, a CPS-improved data estimation model is proposed in this report that can be applied to data anomalies under different meteorological conditions in addition to reliable data estimation. As the core analysis module of the data service framework, the model can be trained based on historical hydrological data and facilitates the detection and estimation of abnormal data, thereby providing reliable data services. The remainder of this paper is organized as follows: Sect. 12.2 discusses the work on data validity in hydrological sensor networks and Sect. 12.3 presents a service framework for the design of flash flood warning sensors, highlighting the design details of the data estimation model. Section 12.4 presents the experimental results for the data estimation model in the actual case, Sect. 12.5 analyzes and discusses the experimental results, and Sect. 12.6 is a summary of this investigation.

12.2   Related Work

This section introduces research related to CPS concerning service architecture design, network communication, and artificial intelligence assisted decision-making of disaster warning systems. In addition, a summary of the achievements in data reliability research of early warning flash flood systems is provided.

12.2.1   Research of CPS in Emergency Service Systems

In various types of disaster warning systems, wireless sensor networks (WSN) that integrate network communication and computing have become popular [11], and real-time monitoring and communication are widely used [12]. The CPS has become more common in the design of information systems for environmental monitoring, natural disasters, and social activities [13].

Researchers [14] designed a popular service architecture of CPS that reliably acquires physical device information and establishes a security monitoring and security constraint mechanism for data analysis to improve system reliability. For a real-time disaster warning system, the investigators [15] proposed an event-driven focused service (EDFS) framework that can be applied to the rapid response of various emergency services at the city level. Several investigators [16] designed a dynamic data-driven application system model for disaster management and adaptively combined real-time data in the disaster management system to ensure continuous stability. A recent study [13] examined the safety mechanism and the CPS security problems caused by defects and vulnerabilities in the network component design and strengthened and supplemented the security design of the CPS architecture to improve the resistance and response capability of the system to emergencies. Several investigators [17] used cloud computing to further enhance the system's ability to cope with disasters and processes. These studies focused on improving the stability and reliability of disaster warning systems from the perspective of service architecture and working mechanisms.

There is a clear trend that artificial intelligence-based approaches have become an important means of improving data aggregation capabilities and data reliability. It has been proposed [18] that machine learning algorithms could be used in WSN to improve information processing and network performance. Investigators [19] have studied the application of several machine learning algorithms for the communication layer in WSN and discussed its specific application. The results of studies on computational intelligence methods in WSN [20] have been reported, including data aggregation, data fusion, and routing and task scheduling. Several researchers [21] have proposed a complementary pairing design strategy based on software-defined networks to improve the adaptability and robustness of sensor networks. All of the aforementioned results can be applied to CPS to improve the security and reliability of the system.

As can be seen from the previous work, the scientific service framework of CPS has essentially been formed. The research on general methods such as obtaining data from physical equipment and relying on scientific methods for transmission and analysis decision-making has shown significant progress. However, in disaster warning systems, there is still much work to be done to provide specific analysis and decision-making services based on actual problems.

12.2.2   Research on Data Reliability in Flash Flood Early Warning Sensor Network

Real-time hydrological data is an important reference for flash flood warning [22]. Due to the complexity of mountainous environments, data from early warning flash flood sensor networks are often abnormal [23, 24], which makes it impossible to perform early warning analysis. At present, researchers have performed a significant amount of research on hydrological data recovery [25]. For example, [26] uses chaos theory to configure artificial neural network models, and [27] used fuzzy rule-based methods. Srivastava et al [28] used multivariate regression and multi-covariate methods. In addition, there are methods based on Bayesian theory [29] and Copula modeling [30], based on ANN and ANFIS model methods [31]. These methods have played an important role in the reconstruction of continuous hydrological data for services such as hydrological observations and drought analysis.

Most of the aforementioned work involves the study of data recovery of a single measurement station and the research methods mostly rely on the construction of hydrological models or the analysis of long-term hydrological datasets. Due to the complex natural environment and geographical features of mountainous regions, the complexity of hydrological models, and the lack of long-term effective hydrological data accumulation, it is difficult to apply these methods directly. Therefore, more specific and practical research is needed in the areas of abnormal data monitoring, effective data extraction, and real-time data services in sensor networks.

12.3 Data Service Framework for Flash Flood Early Warning Sensor Network

12.3.1 Composition of Data Service Framework

The data service framework of the flash flood warning sensor network designed in this paper consists of four modules. As shown in Fig. 12.1, the first is the physical module that corresponds to existing network facilities and physical sensors, including satellite communication and other data transmission and computing resources. The physical module is responsible for the detection and transfer of

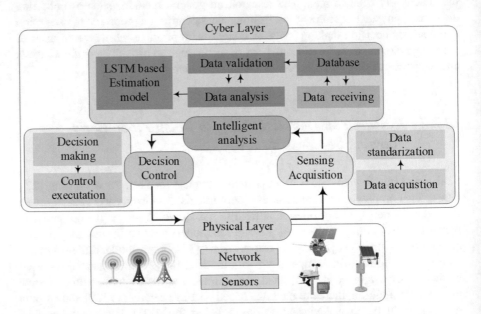

Fig. 12.1 Architecture of CPS-improved data service framework

data of environmental targets. In addition, it is also responsible for the actual implementation and utilization of the decision results. The second is a sensing and acquisition module that is responsible for the acquisition and processing of sensor data. The third is the intelligent analysis module, which can perform data analysis and calculation based on actual data characteristics. The fourth module is the decision and control module, which generates the final decision based on the results of the analysis. As the core of the service framework, the intelligent analysis module not only stores and manages data, but also makes final decisions based on built-in expert modules and analysis modules.

12.3.2   Physical Layer

The measurement station data and weather station data are the main data components in the physical module. The data detected by the sensors are transmitted to the early warning system via the wireless network. The physical layer combines certain historical data records to normalize these data. Because the sensor data varies greatly under different meteorological conditions, the rationality of the data is determined by the analysis algorithm and expert model of the cyber layer.

12.3.3   Cyber Layer

12.3.3.1   Data Sensing and Acquisition Module

The function of this module is to collect, standardize, and manage all types of data required for early warning flash flood analysis. In the working environment, the working state of the gauging sensor is closely related to the meteorological conditions. When rainfall occurs in an area, the acquisition frequency of the gauging sensor increases to improve the ability for real-time determination of a possible flood. When there is no rainfall in the area, the acquisition frequency of the hydrological sensor is significantly reduced to minimize communication costs. Because water flow and temperature changes are directly related in many areas (subtropical and temperate), hydrological data is very sensitive to temperature changes during snowmelt, ice melting, and icing. As such, temperature data are also essential.

In addition, all other data content in the sensor network is managed in this module, including basin information, spatial information, historical hydrology, and meteorological data for each station.

12.3.3.2 Intelligent Analysis Module

The main function of this module is to discover and estimate anomalous data. In this case, a data analysis and estimation model based on LSTM deep learning network is proposed, which can discover data anomalies and use historical data to provide continuous appropriate services. This section focuses on the specific components of this model.

Long short-term memory (LSTM) is a specific recurrent neural network (RNN) architecture introduced by Hochreiter and Schmidhuber [32] and was gradually improved since that time, which is suitable for dealing with problems related to time series. The correlation of the data characteristics of hydrological time series data can be well reflected and applied in this model.

Based on the aforementioned characteristics, the main concept of this model is derived from the LSTM network, based on the analysis of historical hydrological records of the region to determine the data association characteristics of each sensor under different meteorological conditions and use the relevant features identified to evaluate the reliability of each sensor data and provide reliable data estimates.

In order to solve the data relationship characteristics between the sensors under the constraints of the basin, geography and meteorological characteristics, the training and calculation process based on the LSTM network is shown in Fig. 12.2. The properties of each station are also defined in detail: sensor identification number (ID), basin name (B), and geographic location data (G). The observation data input of the model is expressed as vector x: observation time (T), real-time water level data (H), rainfall data (R), and temperature data (T).

A cascade of LSTM layers with n sensors per layer processes information based on the input of the current and of all previous observations. A final softmax layer transforms the LSTM output activation to data predictions.

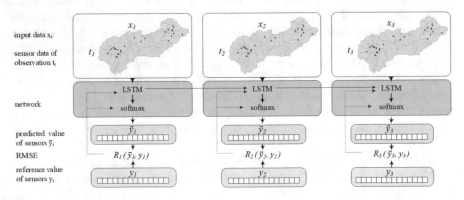

Fig. 12.2 Information flow diagram of our proposed data analysis model based on LSTM network during training and testing

12.3.3.3   Decision and Control Module

As a decision maker in the CPS, the decision and control modules generate the best strategy to meet the needs of the scenario based on the results of the analysis module. In this report, the function of the decision and control modules is based on the analysis results provided by the data estimation model to determine which data to select as the final input to the flash flood warning system.

12.4   Case Study of Data Estimation Model

To verify the data estimation model in the actual environment proposed in this paper, a series of experiments were performed in the Danjiangkou area of Hubei Province, China. The experiments focus on the validity and accuracy of the model under different meteorological conditions.

Danjiangkou City, located in the north-western of Hubei Province, is located between tropic and the temperate zones with an annual rainfall of 750–900 mm. The total area is 3121 km^2. The plain area below 200 m accounts for 18.5%, the hilly area of 200–500 m accounts for 54.9%, the low mountain area of 500–800 m accounts for 19%, and the mountainous area of 800 m or above accounts for 7.6%.

In the experiments, three gauging stations were selected as the research targets, and the complex meteorological scenarios in the study area were divided into groups. Historical hydrological data was used to train the model and each scene was verified using actual data. This report uses TensorFlow-based Keras deep learning framework as background to perform data modeling and statistical processing.

12.4.1   Sensor Network Overview

There are 14 gauging stations and 28 rainfall stations in the study area, of which two have both hydrological and rainfall measurement functions. The data transmission of the entire sensor network relies on wireless communication and the sensor of the gauging station can receive a message to change the data acquisition frequency. According to the hydrological data in recent years, the data abnormality rate of the various sensors is extremely high, in particular, equipment failure and abnormalities occur frequently during the annual precipitation. Due to the complex terrain, there were 26 flash flood evacuation warnings recorded in the system in 2017. Therefore, this study area is suitable for validation of the model proposed in this report.

12.4.2   Dataset of the Experiments

The GuanShan and the JianHe station belong to the same basin but to different river branches; the LangHe belongs to another sub-basin.

The experiments were based on the dataset from 2015 to 2018 which provided nearly 3 years for analysis, training, and prediction to verify the reliability and accuracy of the model. The data for the first two hydrological years (2015.5.22–2017.5.31) was used as a training dataset, accounting for approximately 70

12.4.3   Validation of the Data Estimation Model

Model validation was performed for a complete hydrological year, including all basic types of runoff events that typically occur in the study area. The module validation time was from June 1, 2017, to March 21, 2018. The validation module was developed for the two analyzed sub-basins. This model only verifies the accuracy of data loss in a single gauging station and does not verify data loss from two or more gauging stations.

According to the meteorological characteristics of the study area, four typical scenarios were established to conduct the experiments. All scenarios were based on actual events for the selected time period and all stations have reliable data records covering the entire range of events during verification. The specific scenarios are classified as follows:

S1: Long-term and frequent rainfall. Frequent or continuous regional-scale precipitation occurs frequently, validated from October 5, 2017 to October 18, 2017 and covers the entire time frame before and after the scenario.

S2: Single-peak torrential rain. Typical short-term rainstorms occur frequently; validated from August 3, 2017 to August 13, 2017.

S3: Snowmelt. Typical late-spring snowmelt, with little or no rainfall; validated from February 27, 2018 to March 21, 2018.

S4: Long-term non-rainfall. From June 1, 2017 to March 21, 2018, the model validated the long-term non-rainfall runoff. Low flow is the case of no rain for 18 consecutive days and beyond; validated from January 10, 2018 to January 27, 2018.

Two performance metrics are used in this work to assess the accuracy of the model. The first is the root mean square error (RMSE), which is defined as follows:

$$RMSE = \sqrt{\frac{\sum_{t=1}^{N}\left(x\left(t\right) - \bar{x}\left(t\right)\right)^2}{N}} \tag{12.1}$$

and the other is the mean absolute error (MAE) defined as:

$$MAE = \frac{1}{N}\sum_{t=1}^{N}\left|x(t) - \bar{x}(t)\right| \qquad (12.2)$$

12.4.4 Result

Dataset from June 1, 2017 to March 21, 2018 was used as test dataset to verify the performance of the model in an actual typical complex hydrological year. The validation results show that the overall performance is quite good for a complex hydrological year in the study area, and the (R^2) value ranges from 0.925 to 0.983. The detailed results of the model are shown in Fig. 12.3.

According to the validation results of the GuanShan River Station, the model shows a good fit to the observed value ($R^2 = 0.93$), but the model underestimates the peak flows during the summer and autumn (September 28, 2017 to October 26, 2017) (Fig. 12.3a).

From the validation results of the JianHe station, it is evident that when the water level fluctuation in spring is frequent, the estimation results of the module

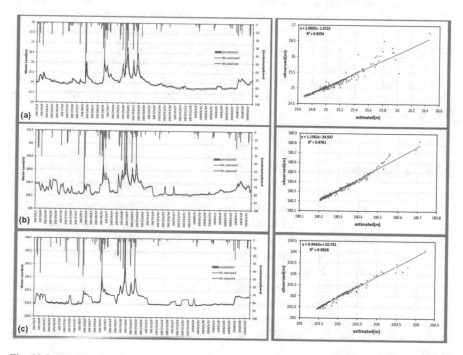

Fig. 12.3 Model validation in a complex hydrological year, June 1, 2017 to March 21, 2018 at stations: (**a**) GuanShan; (**b**) JianHe; and (**c**) LangHe

Table 12.1 Model
performance in the study area

Name	GuanShan	JianHe	LangHe	Mean
Validation	0.9254	0.9781	0.9828	0.9621
S1	0.8495	0.9743	0.9511	0.9250
S2	0.4461	0.9529	0.9278	0.7756
S3	0.937	0.9871	0.9655	0.9632
S4	0.9971	0.8728	0.9995	0.9565
Mean	0.8310	0.9530	0.9653	0.9165

are slightly different, but the overall fitting condition is very good ($R^2 = 0.98$) (Fig. 12.3b).

Among the three sites, the model performed best at the LangHe station ($R^2 = 0.98$). In the spring and summer, the water level is slightly overestimated. However, during the low flow period, the estimates are in good agreement with the observed values (Fig. 12.3c).

The verification of hydrological data in the four scenarios using the model clearly reflects the differences between the different scenarios and the models in the basin (Table 12.1).

The model in the upper part of the basin achieves the best performance, and the performance increases as the average elevation of the basin increases. In the LangHe station located in the highest catchment area, the LSTM network performed well in all scenarios, with R^2 exceeding 0.92 and $R^2_{mean} = 0.965$ for each scenario.

In the middle of the basin, the model exhibited very good performance for the validation of various scenarios in Jianhe station, $R^2_{mean} = 0.953$.

GuanShan, the lowest station of the experiments, has the worst performance for the summer storm scenario. In addition to good performance in low flow scenario, the data estimates in other scenarios are the worst of the three stations with $R^2_{mean} = 0.83$ reliability.

12.5   Discussion

From the aforementioned experimental results, it can be confirmed that the data estimation model based on LSTM performs well in the case of hydrological data anomalies. The comparison of data analysis for different scenarios shows that the model can cope with most meteorological conditions and can deal with extreme conditions such as data loss.

It is evident from the experiment that the model's performance varies under different scenarios. In addition, it can be seen from the results that due to differences in natural geography and environmental conditions, the estimation accuracy of the model is different for sensors in different watersheds or sensors with the same basin but different locations.

Therefore, as the core component of the data service model of the early warning flash flood sensor network, this model demonstrates good data analysis and estimation capability, which can be effectively utilized to address various types of anomalies. Furthermore, as the core module of CPS analysis, the model can simplify the decision process for effectiveness estimation.

The use of this model is limited by the following factors:

(1) Particularity of the model: Models need to select variables and parameter sets for specific basins. Trained models cannot be applied to other basins.
(2) Model prediction period: The experiments performed using this model are for short-term data estimates. For long-term hydrological data estimation, the accuracy of the data will gradually decrease, so the prediction of this model is time-constrained.
(3) Risk of insufficient model training: Reliable models need to be trained using a large amount of data, but due to the limited availability of data for some scenarios, the accuracy of some training results is not good.

12.6   Conclusion

At present, the application of CPS in a sensor-based real-time service system is growing rapidly. It provides a complete framework design and model for investigating comprehensive integration services such as physical facilities, information processing, and intelligent analysis. For the real-time exception handling in a flash flood warning system that was considered in this study, the CPS framework is effective.

Based on the CPS architecture, this report proposes a data service framework that can be used for an early warning flash flood sensor network. The report focuses on the design for data anomaly processing and the estimation model based on the LSTM network and verifies their effectiveness in real scenarios. The results of the experiments prove that the model has good data estimation ability under a variety of meteorological conditions, and can effectively deal with anomalies of the early warning flash flood sensor network while providing stable data services.

The data estimation ability of the model was verified based on the estimation of the daily data and the model was further verified using four typical scenarios: long-term and frequent rainfall, single-peak torrential rain, snowmelt, and long-term non-rainfall. The study shows that the CPS-improved data estimation model can be used as a data validity analysis and calculation module in flash flood warning service. It can handle data anomalies in most scenarios and can simplify the system judgment and decision process.

References

1. E. Gaume, V. Bain, P. Bernardara, O. Newinger, M. Barbuc, A. Bateman, L. Blaškovičová, G. Blöschl, M. Borga, A. Dumitrescu, I. Daliakopoulos, J. Garcia, A. Irimescu, S. Kohnova, A. Koutroulis, L. Marchi, S. Matreata, V. Medina, E. Preciso, D. Sempere-Torres, G. Stancalie, J. Szolgay, I. Tsanis, D. Velasco, A. Viglione, A compilation of data on European flash floods. J. Hydrol. **367**(1-2), 70–78 (2009)
2. S.N. Jonkman, Global perspectives on loss of human life caused by floods. Nat. Hazards **34**(2), 151–175 (2005)
3. D. Liu, S. Zhong, Q. Huang, Study on risk assessment framework for snowmelt flood and hydro-network extraction from watersheds, in *Geo-Informatics in Resource Management and Sustainable Ecosystem* (Springer, Berlin, 2015), pp. 638–651
4. M. Borga, E. Gaume, J.D. Creutin, L. Marchi, Surveying flash floods: gauging the ungauged extremes. Hydrol. Process. **22**(18), 3883–3885 (2008)
5. M.S. Chubey, S. Hathout, Integration of RADARSAT and GIS modelling for estimating future Red River flood risk. GeoJournal **59**(3), 237–246 (2004)
6. A. Bröring, J. Echterhoff, S. Jirka, I. Simonis, T. Everding, C. Stasch, S. Liang, R. Lemmens, New generation sensor web enablement. Sensors **11**(3), 2652–2699 (2011)
7. F. Zhang, S. Zhong, S. Yao, C. Wang, Q. Huang, Ontology-based representation of meteorological disaster system and its application in emergency management: illustration with a simulation case study of comprehensive risk assessment. Kybernetes **45**(5), 798–814 (2016)
8. S. Zhong, Z. Fang, M. Zhu, Q. Huang, A geo-ontology-based approach to decision-making in emergency management of meteorological disasters. Nat. Hazards **89**(2), 531–554 (2017)
9. R. Mariappan, P.V. Narayana Reddy, C. Wu, Cyber physical system using intelligent wireless sensor actuator networks for disaster recovery, in *2015 International Conference on Computational Intelligence and Communication Networks (CICN)* (2015), pp. 95–99
10. R. (Raj) Rajkumar, I. Lee, L. Sha, J. Stankovic, Cyber-physical systems, in *Proceedings of the 47th Design Automation Conference on—DAC'10* (2010), p. 731
11. C.I. Wu, H.Y. Kung, C.H. Chen, L.C. Kuo, An intelligent slope disaster prediction and monitoring system based on WSN and ANP. Expert Syst. Appl. **41**(10), 4554–4562 (2014)
12. C.W. Callaghan, Disaster management, crowdsourced R&D and probabilistic innovation theory: toward real time disaster response capability. Int. J. Disaster Risk Reduct. **17**, 238–250 (2016)
13. A. Mosenia, S. Sur-Kolay, A. Raghunathan, N.K. Jha, DISASTER: dedicated intelligent security attacks on sensor-triggered emergency responses. IEEE Trans. Multi-Scale Comput. Syst. **3**(4), 255–268 (2017)
14. J. Wan, D. Zhang, S. Zhao, L. Yang, J. Lloret, Context-aware vehicular cyber-physical systems with cloud support: architecture, challenges, and solutions. IEEE Commun. Mag. **52**(8), 106–113 (2014)
15. C. Xiao, N. Chen, J. Gong, W. Wang, C. Hu, Z. Chen, Event-driven distributed information resource-focusing service for emergency response in smart city with cyber-physical infrastructures. ISPRS Int. J. Geo Inf. **6**(8), 251 (2017)
16. I. Kureshi, G. Theodoropoulos, E. Mangina, G. O'Hare, J. Roche, Towards an info-symbiotic decision support system for disaster risk management, in *2015 IEEE/ACM 19th International Symposium on Distributed Simulation and Real Time Applications (DS-RT)* (2015), pp. 85–91
17. C.O. Rolim, F.L. Koch, C.B. Westphall, J. Werner, A. Fracalossi, G.S. Salvador, A cloud computing solution for patient's data collection in health care institutions, in *2nd International Conference on eHealth, Telemedicine, and Social Medicine, eTELEMED 2010, Includes MLMB 2010; BUSMMed 2010*, vol. 12(ii) (2010), pp. 95–99
18. M. Di, M.J. Er, A survey of machine learning in wireless sensor networks—from networking and application perspectives, in *2007 6th International Conference on Information, Communications and Signal Processing (ICICS)* (2007), pp. 1–5

19. A. Förster, A. Murphy, Machine learning across the WSN layers, in *Cdn.Intechweb.Org* (2004), pp. 165–183
20. R.V. Kulkarni, A. Förster, G.K. Venayagamoorthy, Computational intelligence in wireless sensor networks: a survey. IEEE Commun. Surv. Tutorials **13**(1), 68–96 (2011)
21. R.W. Skowyra, A. Lapets, A. Bestavros, A. Kfoury, Verifiably-safe software-defined networks for CPS, in *Proceedings of the 2nd ACM International Conference on High Confidence Networked Systems (HiCoNS '13)* (ACM, New York, 2013), p. 101
22. S. Zhu, J. Zhou, L. Ye, et al. Streamflow estimation by support vector machine coupled with different methods of time series decomposition in the upper reaches of Yangtze River, Environ. Ear. Sci. China [J], 2016, **75**(6), p. 531
23. J. Langhammer, J. Česák, Applicability of a nu-support vector regression model for the completion of missing data in hydrological time series. Water (Switzerland) **8**(12), 560 (2016)
24. A.J. Rettig, S. Khanna, D. Heintzelman, R.A. Beck, An open source software approach to geospatial sensor network standardization for urban runoff. Comput. Environ. Urban. Syst. **48**, 28–34 (2014)
25. P. Coulibaly, N.D. Evora, Comparison of neural network methods for infilling missing daily weather records. J. Hydrol. **341**(1-2), 27–41 (2007)
26. A. Elshorbagy, S.P. Simonovic, U.S. Panu, Estimation of missing streamflow data using principles of chaos theory. J. Hydrol. **255**(1-4), 123–133 (2002)
27. A.J. Abebe, D.P. Solomatine, R.G.W. Venneker, Application of adaptive fuzzy rule-based models for reconstruction of missing precipitation events. Hydrol. Sci. J. **45**(3), 425–436 (2000)
28. A.K. Srivastava, M. Rajeevan, S.R. Kshirsagar, Development of a high resolution daily gridded temperature data set (1969–2005) for the Indian region. Atmos. Sci. Lett. **10**(October), 249–254 (2009)
29. M.P. Tingley, P. Huybers, A Bayesian algorithm for reconstructing climate anomalies in space and time. Part I: Development and applications to paleoclimate reconstruction problems. J. Climate **23**(10), 2759–2781 (2010)
30. A. Bárdossy, G. Pegram, Infilling missing precipitation records—a comparison of a new copula-based method with other techniques. J. Hydrol. **519**(PA), 1162–1170 (2014)
31. M.T. Dastorani, A. Moghadamnia, J. Piri, M. Rico-Ramirez, Application of ANN and ANFIS models for reconstructing missing flow data. Environ. Monit. Assess. **166**(1–4), 421–434 (2010)
32. S. Hochreiter, J. Schmidhuber, Long short-term memory. Neural Comput. **9**(8), 1735–1780 (1997)

Index

© Springer Nature Switzerland AG 2020
S. Hu, B. Yu (eds.), *Big Data Analytics for Cyber-Physical Systems*,
https://doi.org/10.1007/978-3-030-43494-6

Printed in the United States
by Baker & Taylor Publisher Services